Stormwater Manufactured Treatment Devices

Certification Guidelines

Joint Task Committee on Guidelines
for Certification of Manufactured Stormwater BMPs

Edited by
Qizhong Guo, Ph.D., P.E., D.WRE

Sponsored by the
Sustainable Stormwater Infrastructure Committee of the
Water, Wastewater, and Stormwater Council and the
Wet Weather Flow Technologies Committee of the
Urban Water Resources Research Council of the
Environmental and Water Resources Institute of the
American Society of Civil Engineers

Published by the American Society of Civil Engineers

Library of Congress Cataloging-in-Publication Data

Names: Joint Task Committee for Certification of Manufactured Stormwater BMPs, author. | Guo, Qizhong, editor.
Title: Stormwater manufactured treatment devices certification guidelines / Joint Task Committee on Guidelines for Certification of Manufactured Stormwater BMPs ; edited by Qizhong Guo, Ph.D., P.E., D. WRE ; sponsored by the Sustainable Stormwater Infrastructure Committee of the Water, Wastewater, and Stormwater Council and the Wet Weather Flow Technologies Committee of the Urban Water Resources Research Council of the Environmental and Water Resources Institute of the American Society of Civil Engineers.
Description: Reston, Virginia : American Society of Civil Engineers, 2017. | Includes bibliographical references and index.
Identifiers: LCCN 2017032835| ISBN 9780784414798 (soft cover : alk. paper) | ISBN 9780784480670 (PDF) | ISBN 9780784480687 (ePUB)
Subjects: LCSH: Urban runoff–Management–Standards.
Classification: LCC TD657 .J65 2017 | DDC 628/.210284–dc23 LC record available at https://lccn.loc.gov/2017032835

Published by American Society of Civil Engineers
1801 Alexander Bell Drive
Reston, Virginia 20191-4382
www.asce.org/bookstore | ascelibrary.org

Errata: Errata, if any, can be found at https://doi.org/10.1061/9780784414798.

ISBN 978-0-7844-1479-8 (print)
ISBN 978-0-7844-8067-0 (PDF)
ISBN 978-0-7844-8068-7 (ePUB)
Manufactured in the United States of America.

24 23 22 21 20 19 18 17 1 2 3 4 5

Contents

Acknowledgments

JOINT TASK COMMITTEE ON GUIDELINES FOR CERTIFICATION OF MANUFACTURED STORMWATER BEST MANAGEMENT PRACTICES

Officers

Chair:	Qizhong (George) Guo
Vice Chair:	Gordon England
Secretary:	Charlene Johnston

Chairs of the Subcommittees

Laboratory Testing:	Omid Mohseni
Physical Scaling:	John Gulliver
Field Testing:	John Sansalone
Data Evaluation:	Masoud Kayhanian
Data Reporting:	Rob Roseen
Maintenance:	Bill Hunt

Liaisons

WERF:	Jim Lenhart
ASTM:	Ernie Carrasco

Subcommittees

Laboratory Testing

Omid Mohseni (Chair), Shohreh Karimipour, Jon McDonald, Hans de Bruijn, Shirley Clark, Scott Perry, Kwabena Osei, Mark Miller, Fariar Kohzad, Roger Bannerman, Jim Mailloux, Masoud Kayhanian, Daniel Smith

Physical Scaling

John Gulliver (Chair), John Sansalone, Kwabena Osei, George Guo, Jim Mailloux, Gregory Williams, Daniel Smith, Jy S. Wu

Field Testing

John Sansalone (Chair), Betty Rushton, John Gulliver, Rob Roseen, Uday Khambhammettu, Jeff Benty, Bill Hunt, Ernie Carrasco, Jon Hathaway

Data Evaluation

Masoud Kayhanian (Chair), Shirley Clark, Shohreh Karimipour, Jon McDonald, Rob Roseen, Uday Khambhammettu, Ernie Carrasco, Jim Lenhart, Jim Mailloux, Gregory Williams

Data Reporting

Rob Roseen (Chair), Charlene Johnston, Walt Stein, Bill Hunt, Yuan Cheng, Jim Mailloux

Maintenance

Bill Hunt (Chair), Omid Mohseni, Hans de Bruijn, Scott Perry, Mark Miller, Gordon England, George Guo, Richard Gee

Full Task Committee

Nicholas Agnoli, Mathew Anderson, Lily S. Baldwin, Roger Bannerman, Michael Barrett, Jeffrey Benty, Graham Bryant, Michael T. Buechter, Ernie Carrasco, Yuan Cheng, Fred Civian, Shirley E. Clark, Mathew D. Davis, Hans DeBruijn, Thomas R. Decker, Gordon England, Elizabeth Fassman, Richard Field, Ray M. Frederick, Richard Gee, Qizhong (George) Guo, John S. Gulliver, Jon Hathaway, Raymond Hozalski, William F. Hunt, Tom Jacobs, Charlene E. Johnston, Masoud Kayhanian, Shohreh Karimipour, Jong-Yeop Kim, Kevin Kirsch, Uday Khambhammettu, Fariar Kohzad, James H. Lenhart, James T. Mailloux, Tom Maguire, Jonathan McDonald, Mark B. Miller, Omid Mohseni, Kenneth Murin, Kwabena Osei, Linda Pechacek, Scott Perry, Ravi Patraju, Robert Pitt, Sri Rangarajan, Larry A. Roesner, Robert M. Roseen, Betty Rushton, David J. Sample, John J. Sansalone, Daniel P. Smith, Stuart Stein, Walt Stein, Tom Stevens, Scott Struck, Robert G. Traver, Don Wade, Gregory Williams, Tim Williams, Grace Wong, Len Wright, Jy S. Wu

Full Committee Report Workgroup

Qizhong (George) Guo, Gordon England, Charlene Johnston, Jim Lenhart, Bill Hunt

Review Panel

Richard Field, Tom Maguire, Robert Pitt, Larry A. Roesner

Preface

Since the passage of the U.S. Clean Water Act in 1972 and subsequently through the National Pollution Discharge Elimination System (NPDES) amendment in 1983, many local, state, and industrial entities are obligated to treat stormwater runoff before discharge to receiving waters. Initially with NPDES Phase I permits, regulatory compliance was only applicable to communities with a population higher than 100,000. Since 2002, NPDES Phase II permits extended the same regulatory framework to communities with populations of 10,000 or more and to areas along the urban fringe that together have a residential population of at least 50,000 and an overall population density of at least 1,000 people per square mile. For this reason, demand for the use of compact stormwater treatment systems has increased rapidly where land space is not readily available to accommodate other types of systems. Numerous manufacturers have responded and developed a wide range of structural best management practices (BMPs), now referred to as stormwater control measures (SCMs) to remove different pollutants. For purposes of this report, these compact stormwater treatment systems are called manufactured treatment devices (MTDs).

Most MTDs function either as hydrodynamic separators (HDSs) or as filters. HDSs utilize the difference in density (between the water and stormwaterborne particles) and flow patterns to separate by inertia and/or settle by gravity stormwaterborne particles to the bottom of the device and to retain floatables, such as hydrocarbons and trash, at the surface of the device, while the effluent flows through an intermediate elevation. Filters use a medium to retain particles of a given size while the effluent flows through the filter. These devices are generally placed below ground in high population density locations and retrofits to existing infrastructure.

In recent years, the use of MTDs has increased rapidly in response to escalating water quality regulatory requirements, particularly in redevelopment and new development areas where land space is limited and potential use of other SCMs is not practical. As the use of MTDs has increased, the demand for verification of and certification for pollutant removal and meeting water quality standards has also increased. Agencies have started to realize that without a program, the process of approving, sizing, and providing regulatory oversight of MTDs is problematic. However, the establishment of a program is time consuming and expensive and requires multidisciplinary expertise. While several verification and certification programs have been proposed at the federal, state, and regional levels (e.g., TAPE, TARP, and ETV, as discussed in Chapter 7), there is no consistent protocol at a national level.

These Guidelines provide a framework to create verification and/or certification programs across a broad range of conditions. A verification process is a means by which an independent party of reviewers conducts a thorough review of lab and/or field data reports. The review verifies that means of data collection, reporting, and statements are correct and in accordance with a preapproved project quality assurance project plan (QAPP). A certification, within the context of stormwater treatment, is a process through which a regulatory agency reviews the verified documents, determines whether the SCM should be approved or not for use within its territory, and provides guidelines for sizing and design that are unique to the agency's jurisdictional area. Certifications are more comprehensive than verifications.

The development of a framework that can be used across the country has the potential of providing comparable data across regulatory agencies and regions, thus facilitating the certification process in the long run. The Guidelines for the Certification of Manufactured Stormwater BMPs (Guidelines) task committee was established under the Environmental and Water Resources Institute's (EWRI's) Sustainable Stormwater Infrastructure Committee of the Water, Wastewater, and Stormwater Council and the Wet Weather Flow Technologies Committee of the Urban Water Resources Research Council. In addition, parallel efforts are being undertaken by ASTM International and Water Environment Research Foundation (WERF). To avoid inconsistencies among the organizations, liaisons from the task committee were established for these other groups. To develop the certification guidelines, several subcommittees were established and each subcommittee was responsible for the development of one of the following components: laboratory testing, scaling relations, field testing, data reporting, data analysis and performance evaluation, and maintenance. The practitioner is reminded that the material presented herein is for guidance purposes only. Organizations and jurisdictions may require specific procedures to be followed that are not outlined herein. Irrespective of procedure used, it must be clearly outlined in the QAPP.

This effort has been broadened to be consistent with the reporting requirements for the International Stormwater BMP Database (BMPDB) developed by a coalition of partners including WERF, ASCE EWRI, U.S. EPA, Federal Highway Administration, and the American Public Works Association. The purpose of the BMPDB is to provide scientifically sound information to improve the design, selection, and performance of SCMs.

For a community to establish a successful stormwater management program there are many considerations, including defining realistic regulations, understanding the intangible elements of managing a successful SCM program, understanding specific MTD design requirements, and ensuring designs and installations are completed in accordance with the design specifications. Sufficient budgeting is critical to the long-term success of any program. Continuing education and training for regulatory staff are necessary to keep up with the constant changes in regulations, technology, and methods to measure and evaluate the technologies.

PURPOSE AND SCOPE

This document has two primary purposes. The first is to assist regulators and end users in assessing the efficiency of MTDs (HDSs and filters) used for physical separation of the stormwater runoff particulate fraction (the full spectrum of stormwaterborne nondissolved solids) from stormwater runoff liquid (water) fraction. The Guidelines are intended to include the assessment of scouring of settled particles under high flow conditions. The Guidelines do not include removal of trash, leaves, and hydrocarbon products.

The information generated using these Guidelines can be incorporated into models to better predict the net annual device efficiency when deployed in a specific location under specified rainfall patterns and pollutant characteristics.

The second purpose of these Guidelines is to aid regulators in establishing certification programs for these devices. These Guidelines are intended to provide a broad set of analytical tools that can be adapted to the particular characteristics of a regulatory agency's jurisdiction. Programs based on consistent guidelines across diverse regions can provide advantages both for regulatory agencies and manufacturers. Regulatory agencies can use these Guidelines to adapt design standards suitable for their specific locations and needs. Consistent testing and analytical methodologies among communities will enable manufacturers to focus limited resources on improving product designs to meet a limited number of design standards and certifications.

These Guidelines address seven critical components that should be used in the development of design criteria, testing, verification, and certification of MTDs:

1. Laboratory Testing
2. Scaling Relations
3. Field Testing
4. Data Analysis and Performance Evaluation
5. Data Reporting
6. Inspection and Maintenance
7. Verification and Certification Programs and Processes

In addition, samples of certification programs used in various regions are briefly discussed.

Qizhong Guo, Ph.D., P.E., D.WRE
Chair

Terms and Definitions

Laboratory Testing Subcommittee

Accuracy: The degree of agreement between the quantity measured and the true quantity.

Average Velocity: The average velocity of a stream flowing in a channel or conduit at a given cross section or in a given reach is equal to the flow rate divided by the cross-sectional area of the section or the average cross-sectional area of the reach.

Background Concentration: Background concentrations in a closed loop system are solids that are in the influent stream prior to the feeder adding solids to the influent stream, quantified as concentration (mg/L) or mass per unit volume of water.

Baffles: Deflector guides, grids, gratings, or similar devices constructed or placed in flowing water to absorb energy; divert, guide, or agitate the liquids; and check eddies.

*Bed Load: S*ediment that moves by saltating, rolling, sliding, or skipping along the bed or essentially in contact with the invert of the conveyance structure.

Bypass Flow: Flow that is designed to bypass the treatment components of the stormwater control practice. Untreated flows less than the design flow are termed overflow. In some instances, the overflow may be directed through the bypass flow route or commingled with the bypass flow.

Calibration: The obtaining of quantitative measurements to certify the measurements obtained by any instrument, such as a flow meter or rain gauge.

Discharge: The volume of water that passes through a given location within a given period of time.

Flow Rate: Same as *Discharge.*

Head Loss: The decrease in total energy of moving water between the entrance and exit points of a control practice. One common type of head loss is the frictional head loss due to friction between flowing water and the wall or surface of the conduit, channel, and/or device. Local losses are due to abrupt changes in conduit geometry, such as bends, valves, and shifts in diameter and width.

Hydrodynamic Separator (HDS): Flow-through devices used as pretreatment in a multitreatment stormwater control measure (SCM) train or as standalone SCMs. Used for removing contaminants from stormwater. Water either enters these devices tangentially, thus creating a swirl or vortex, or plunges into the main sump. Hydrodynamic separators may be single or multiple sump devices.

Invert: Elevation at the inside bottom of the pipe, conduit, or channel.

Manufactured Treatment Device (MTD): A manufactured device for the removal of pollutants from stormwater runoff.

Maximum Design Treatment Rate (MDTR): The maximum flow rate (volume/time) used to design the treatment device for the purposes of treatment.

Maximum Hydraulic Capacity: Maximum flow rate accommodated within an acceptable head loss limit. This includes the MDTR plus the bypass capacity.

Maximum Storage Depth: This is the maximum depth of sediment accumulation recommended by the manufacturer to maintain acceptable sediment removal efficiency and scour losses. This depth can be measured from the invert of the inlet pipe or as the depth from the bottom of a collection sump.

Media Filtration: A unit process that removes particles and associated pollutants by directing stormwater under gravity or pressure through a media filter. Sedimentation is also a unit process on the surface of the media or within the media. Media filtration may also be used to remove dissolved constituents, such as metals and nutrients. The processes of surface complexation, sorption, precipitation, and ion exchange in a medium remove dissolved constituents as well. The processes involved will depend on the selection of medium.

Off-Line Devices: Control practices that only receive a portion of the water generated by the drainage area, especially during larger events. Stormwater is delivered to the practice to a prespecified maximum flow rate or volume by an upstream flow diversion structure. A practice that is off-line can be made smaller, with the upstream flow diversion functioning as a bypass to reduce the chance of scour and improve overall net performance.

Online Devices: A control practice designed to receive all the flow from the drainage area, including large events.

Particle Size Distribution (PSD): A description of the range of different size particles in a sediment sample as a fraction of mass of the total sample.

Physical Separation: Sediment particles being removed or separated from water by gravity and/or inertial forces and/or filtration.

Scour: Scour is the resuspension of previously captured sediments. Scour is caused by excessive shear stress from flows reaching the deposited sediments in the sediment storage chamber or region. Scour of particles may not necessarily be associated with washout in that scouring may result only in the movement or resurfacing of a sediment pile profile.

Sedimentation (Gravitational Separation): Sedimentation is a unit process that utilizes gravity to remove settleable particles and associated pollutants from aqueous solution. The difference in particle and water densities is integral to sedimentation or surface loading rates.

Slope: The rate of ascent or descent expressed as a percentage determined by the ratio of the change in elevation to a defined length.

Specific Gravity: The ratio of the mass of a body to the mass of an equal volume of water at 68°F.

Stokes' Law: Describes the settling velocity of spherical particles. The equation for Stokes' law illustrates the relationship of various factors that affect the settling velocity of particles in a water column when the grain Reynolds number is less than 3. It is suitable to estimate settling velocity of spherical particles roughly smaller than 100 μm.

Surface Loading Rate: The flow rate (Q) in volume per unit time divided by the surface area of a settling, filtration, or treatment device (A).

Suspended Sediment Concentration (SSC): The measure of the dry weight of sediment from a known volume of a water-sediment mixture. Determined analytically by test method ASTM D3977-97.

Suspended Solids: Solids suspended in stormwater due to their small size and molecular weight or to the motion of the water. The definition does not include gross solids, such as gravel and litter. For purposes of these Guidelines, suspended solids can be organic and/or inorganic material.

Total Suspended Solids (TSS): The measure of the dry weight of sediment present in a known volume of a sub-sample of the original sample. It is also referred to as total nonfilterable residue. Determined analytically by Standard Method 2540-D (APWA, AWWA, WEF 2012).

Treatment (Performance) Efficiency: Treatment efficiency is the reduction in pollutant concentrations or loads between the inlet and the outlet of the stormwater treatment device or percent reduction based on mass entering the device and mass retained.

Turbidity: Optical property of a sample that causes light rays to be scattered and absorbed; this parameter is not an accurate measure of the mass or concentration of sediment in the sample.

Velocity: The rate of motion of water or particles in distance per unit time in a specific direction.

Washout: Washout occurs when previously settled sediments are scoured and transported out of the treatment unit through the outlet.

Weir: An artificial control structure with a rating curve that can be used for measuring flow in a conduit or chamber for diverting design flow.

CHAPTER 1

Laboratory Testing

Laboratory Testing Subcommittee

Laboratory testing allows for a device to be tested under controlled conditions while varying only select parameters such as flow rate, concentration, or particle size/settling velocity distribution. This level of control, while not wholly representative of actual field conditions, serves as a good indicator of the expected performance of the device under field conditions. It also allows for product improvement by varying configurations under controlled conditions and direct comparison from one type of device to another.

A significant amount of work has been accomplished by ASTM International (ASTM) C27.70 in the development of detailed laboratory procedures. The work items completed are listed in the references (ASTM 2011, ASTM 2012, ASTM 2016).

To provide clear guidelines on the most suitable protocol for obtaining accurate, reliable, and replicable results, this section on laboratory testing has been divided into five subsections. The first subsection outlines the objectives of performing laboratory testing. The second subsection outlines the qualifications that the laboratory and data analysts should have to perform the testing. The third subsection gives guidelines on the procedure to test the removal efficiency of hydrodynamic separators (HDSs). The fourth subsection gives guidelines for testing the sediment retention capabilities of HDSs. The final subsection deals with testing of filtration devices. Special emphasis has been placed on providing descriptions of the recommended test materials, hydraulic characteristics, removal rate measurements, and measurement of scour and washout in a preloaded device.

Even though most certification programs currently being implemented use total suspended solids (TSS) removal efficiency, these Guidelines recommend using a mass balance approach for laboratory testing. TSS was originally developed for wastewater testing; recent testing has shown that TSS produces inaccurate results for stormwater removal efficiency (Roesner et al. 2007). In some cases, it might be acceptable to use suspended sediment concentration (SSC) removal efficiency as described further in the following sections. It should be noted that other improved analytical methods have been proposed by Roesner (2007) for testing the removal efficiency of stormwater manufactured treatment devices (MTDs); however, until these methods have been validated and standardized, it is recommended to use methods for which standard procedures are available and easily replicable.

1.1 LABORATORY TESTING OBJECTIVES

- To quantify the mass of solid particles trapped by a device under different flow rates in a laboratory setting;
- To present and analyze data to show device efficiency as a function of particle size, particle settling velocity, flow rate, and other physical characteristics;
- To quantify effluent concentration resulting from scouring, resuspension, and washout of previously deposited sediment under different flow rates; and
- To demonstrate scaling relationships for predicting the efficiency of the device for untested particle sizes, specific gravities, and flow conditions.

1.2 LABORATORY AND DATA ANALYST QUALIFICATIONS

1.2.1 Laboratory Qualifications

- An independent laboratory shall conduct MTD laboratory testing. Whenever the testing is performed at the manufacturer's own laboratory, it shall be overseen by an independent party.
- The laboratory conducting the performance testing must be able to provide the range of flows, sediment characteristics, measurement and recording systems, and trained personnel necessary to generate reliable test results. A general statement of laboratory qualifications shall be submitted with the report.
- If the manufacturer is using its own laboratory and an independent observer, the observer shall meet the following requirements:
  - The observer shall have no financial or personal conflict of interest regarding the test results.
  - The observer shall have experience in hydraulics and sedimentation, be familiar with the test and laboratory methods specified in this standard, and have a professional license in an appropriate discipline. A statement of the qualifications of the observer shall be submitted with the report.
  - The observer shall approve the experimental setup and laboratory testing protocol, observe the test during its full duration, and sign with a professional seal a statement that verifies the completeness and accuracy of the report.

1.2.2 Data Analyst Qualifications

- The analysis of laboratory data shall be performed by a qualified individual. A statement of qualifications for the selected individual shall be submitted with the report.
- The qualifications for the data analyst shall be reviewed and approved by the verification entity.

1.3 REMOVAL EFFICIENCY TESTING OF HYDRODYNAMIC SEPARATORS

A mass balance of sediment entering the device and retained in the retention chamber provides the MTD removal efficiency. The default method of removal efficiency testing of these Guidelines is to measure the mass of sediment fed into the influent pipe and the mass retained in the device. SSC shall be used to measure the influent and effluent MTD concentrations. If the laboratory shows that it can sample the influent SSC with accuracy better than 90% for the chosen ranges of particle sizes, then the SSC of the influent and effluent can be used to determine the efficiency of the device. This can be accomplished by taking flow-weighted samples of the influent and effluent and calculating the total mass in and total mass out. The result should be within 10% of the actual measured mass in and the difference between the actual mass in minus the actual mass retained; otherwise the tests need to be rerun.

The HDS performance is evaluated by measuring the removal efficiency of a specified particle size distribution (PSD) using five different flow rates.

1.3.1 Testing Objective

To quantify the removal efficiency of silica sediment under varied conditions of flow and particle size.

1.3.2 Testing Parameters and Procedure

1.3.2.1 Testing Parameters

Determining Maximum Design Treatment Rate: The maximum design treatment rate (MDTR) is widely used in the industry but has several definitions. One definition is that rate at which bypassing begins, others use the point at which the treatment efficiency for a given particle size is zero. Still others (Lenhart et al. 2009) suggest this point be at the optimum operating point.

Clearly, if the treatment efficiency is zero and perhaps on the brink of resuspension, this should not be the MDTR. In addition, because the recommendations for bypassing vary by manufacturer and other conditions, the MDTR becomes a variable. Because the optimum treatment rate can be calculated, the MDTR is defined to be the optimum operating rate for the purposes of these Guidelines.

Flow Rates: At a minimum, each device shall be tested at the following discrete steady-state flow rates:

- 25% of the MDTR,
- 50% of the MDTR,
- 75% of the MDTR,
- 100% of the MDTR,
- 125% of the MDTR, and

- If different from above, the *optimum design treatment rate* (Lenhart et al. 2009).

Test Sediment: A minimum of three discrete particle size ranges (e.g., 63 to 68 μm, 149 to 210 μm, and 354 to 420 μm) will be used for testing. The test results from these distinct size ranges can then be used to compose and represent local site conditions. Sediment particle size ranges should be selected to represent particle sizes in the vicinity of the region where certification is desired.

The sediment will be completely mixed prior to passing through the device. Sediment injection directly at the inlet into the device itself shall not be allowed to avoid errors in the results from incomplete mixing or bypass of sediment due to its concentration in only part of the test water. The approach velocities in the inflow pipe should be consistent with the maximum design velocities in the field.

Influent Concentration: The practitioner must demonstrate influent SSC concentrations of 200 ±40 mg/L at the beginning, halfway point, and near the end of each test run. The quality assurance project plan (QAPP) must include the procedure using either a wet or dry technique.

Supply Hopper: The hopper shall be calibrated to dispense a consistent amount of test sediment.

Water Temperature: At a minimum, water temperature shall be measured at the beginning, halfway point, and near the end of each test run. The average of these measurements will be used as average water temperature for calculations and reporting.

1.3.2.2 Testing Procedure

- Practitioners are advised to utilize the standards developed by ASTM subcommittee C27-70. These will be detailed laboratory standards for MTDs.
- The device will be set up with all components and connections required during normal operation.
- The device must be tested starting at 50% of the maximum sediment storage depth. This requirement can be fulfilled by constructing a platform at the 50% maximum storage depth. The platform shall be completely sealed to prevent any particles from passing into the lower level.
- At a minimum, each device will be tested at the five flow rates provided in Section 1.3.2.1.
- Each test shall be performed at least three times to ensure the accuracy of the results.
- Each test shall be run until at least 5 lb of material has been trapped in the MTD.
- At least five one liter samples of water shall be taken at different times during the test to obtain the removal efficiency of the device under different load conditions. Table 1.1 summarizes the recommended sampling procedure.
- SSC will be measured using standard ASTM method D3977-97 (2007).

Table 1.1. Recommended sampling for sediment removal performance test

Sampling Location	*Total Sediment Mass*	*Suspended Sediment Concentration*
Sediment Supply Hopper	Total mass weighed at beginning and end of test	N.A.
Settling Chamber	Total mass collected	N.A.
Effluent Pipe	N.A.	5 samples

- Flow rate shall be monitored throughout the test using a calibrated flowmeter. It is recommended that the procedure used complies with ASTM flow measurement standards (ASTM 2011).

1.4 SEDIMENT RETENTION/SCOUR TESTING OF HDSs

The purpose of this testing method guidance is to help regulators and end users determine and understand the performance of an HDS in a lab setting. Data from this lab testing can be used to characterize performance of these devices under flow conditions in excess of the MDTR and can be incorporated into models to better predict the net annual device efficiency when deployed in a specific location under a specified climate condition and suspended pollutant characteristics.

1.4.1 Testing Objectives

- To quantify effluent concentration (or mass loss) from scouring, resuspension, and washout of previously deposited sediment in an online MTD under different flow conditions.
- To develop a maintenance schedule for the device.

1.4.2 Testing Parameters and Procedure

1.4.2.1 Testing Parameters

- A flow meter (e.g., orifice meter, weir) should be installed either on the influent pipe or the effluent pipe to measure discharge through the device with an accuracy of plus or minus 3% of actual flow. The flowmeter should be installed according to ASTM C27.70 (2013) standards.
- The temperature shall be measured and reported at the beginning, middle, and end of each test.
- Water levels in the influent and effluent pipes and inside the device compartments should be measured and reported from a single datum and

reported along with the flow readings at the time the measurements were taken.

- If the method to measure sediment washout is based on measuring the change in the weight of deposit, load cells with a repeatability of better than 0.05% of maximum capacity of load cells should be used.
- Sediments used for this test might be commercially available but shall have a narrow PSD, e.g., the F110 Silica Sand gradation (i.e., settling velocities comparable to coarse silt or sand particles). Prior to each test, samples of the sediments placed in the sump should be sieved and analyzed per ASTM C27-70 standards (2013).

1.4.2.2 Testing Procedure

- To determine the sediment retention of an HDS, at least two tests per each of five flow rates (from 50% of MDTR to the maximum hydraulic rate) for each PSD need to be run. This is equivalent to a minimum of 30 runs. Tests should be conducted when the deposit in the sump is at 50 and 100% of the maximum storage depth of the device. If the difference between the results of two repeated tests is more than 10%, assuming the discharge and water temperature of the two repeated tests are within 3% of each other, both tests should be repeated.
- The PSD should not contain any particle larger than 600 μm. To mimic the bedforms and shape of the deposit in the sump, the sediments shall be fed into the system using a sediment feeder or slurry and at a water discharge no more than 10% of MDTR. If the testing PSD contains particles smaller than 100 μm, the sediments shall be placed manually inside the device (to avoid error induced by the smaller particles bypassing the device) and then the surface of the deposit shall be leveled.
- Testing for resuspension can start from the maximum hydraulic rate of 50% of MDTR. If two repeated tests at a given flow rate on a specific PSD do not result in any significant effluent concentration due to washout, i.e., less than 25 mg/L or a concentration less than or equal to that of the influent, the tests at lower flow rates on the PSD and deposit level are not required.
- Prior to each test, the sump of the device will be filled with the sediments described previously to either 50 or 100% of the maximum storage depth. For tests where 100% of the maximum storage depth is filled with sediments, a false floor may be utilized to decrease the amount of sediments deposited in the sump prior to the test. The false floor shall be sealed to prevent particles passing underneath the false floor. The depth of the sediment shall be such that the surface of the false floor is at least 0.5 in. below the surface of the sediment.
- A short duration test at discharges less than MDTR may result in large errors, depending upon the amount of washout. Therefore, the test duration should

be long enough to minimize any errors in the measurement of the washed out sediments.

- At the end of each test, the remaining deposit in the sump will be mixed and five samples will be collected. The samples will be sieved. If the mean PSD of the samples are within the variability of the samples taken prior to each test, the deposit in the sump can be replenished to the desired target using the original particle size distribution, otherwise, the sump should be completely cleaned and the original sediment placed in the sump.
- If a recirculating water system is used for the testing, the concentration and the PSD of recirculated water must be known. Samples should be taken immediately before re-entering the MTD.
- If the water supply is not potable, a background concentration test shall be performed. The background concentration test shall consist of running water through the device at 10% of MDTR without feeding any solids into the system for a period of at least one hour. After the 1-h period, the device shall be drained and the solids removed by the device shall be collected, dried, sieved, and weighed. The background concentration for each particle size shall then be estimated using the equation

$$C_{\text{background}} = \frac{W}{Qt}$$

where $C_{\text{background}}$ = background concentration; t = duration (1 h); W = weight of the measured solids; and Q = flow rate set at 10% of the maximum tested flow rate. The weight of the particles that might settle during sediment retention testing shall be estimated using the background concentration and/or the performance function obtained from the removal efficiency testing procedure. The resulting weight shall be added to the initial weight of the deposit in the retention testing.

- After a sediment retention test, if the sump floor is exposed, the result of that test is not suitable for analysis and shall be invalid, because the washout would be limited to the availability of the sediment in the sump. Under such circumstances, the deposit in the sump shall be replenished and the test duration or discharge shall be decreased to avoid any exposure of the sump floor.
- Sediment retention testing can be conducted based on either measuring the volume or weight of washed out sediments.
    - *Measuring Weight.* This method is used to measure the weight of the manhole, the internal components of the device, a volume of water inside the device, and the sediments before and after each test using load cells. The difference between the two weights gives the mass of sediments washed out of the device during each sediment retention test. Given the limited accuracy of load cells, the errors in this method can be as large as 7 kg for a

1.5-m diameter device, which is approximately 35 mg/L at a discharge of 5.7 L/s for a test duration of one hour. As the test duration and discharge increase, the error in this method decreases.

This method can be applied to devices where the total weight of the manhole, the volume of water below the outlet invert, and sediment inside the manhole does not exceed 11,000 kg.

In this approach, the weight of the device, the sediments placed in the sump, and the volume of water at some predefined level in the sump above the deposit is measured prior to each test. The water level shall be measured with an accuracy of 3 mm. Note that 3 mm of water in a 1.5-m diameter sump weighs 2.2 kg. After running water through the device for a defined period of time (e.g., one hour), the sump is drained to the predefined level and the weight of the system is measured again. The water level in the sump shall be measured again, and if it is different from the predefined level, the weight of the system should be adjusted accordingly. Effluent concentration of the washed out sediments will be estimated using the following equation:

$$C_{\text{eff}} = \frac{\Delta W \text{SG}}{Q \Delta t (\text{SG} - 1)}$$

where ΔW = difference between the two weights prior to and after the test; Q = discharge; Δt = duration of the test at the target discharge; and SG = specific gravity of sediments or synthetic particles.

- *Measuring Volume.* This method is used to measure the volume of wet sediments in the sump before and after each test. By employing the wet bulk specific weight of sediments, the change in volume can be converted into change in weight of sediments after each sediment retention test. The error in this method can be as large as 7 kg for a 1.5-m diameter device, which is about 35 mg/L at a discharge of 5.7 L/s for a test duration of one hour. As the test duration and discharge increase, the error in this method decreases.

  This method can be applied to devices that are too heavy for using load cells but whose sumps can be easily accessed by the testing personnel (i.e., the sump is not cluttered with the internal components of the device). To employ this approach, a repeatable test method shall be employed to measure the in situ wet bulk specific gravity of the sediments placed in the sump. The repeatability of the tests to determine the wet bulk specific gravity of sediments shall be within 5%.

  In this approach, after placing the sediments or the synthetic particles in the sump, the sump is filled with water to about 1 cm above the deposit. Then, the sediments in the sump should be flattened and leveled using a trowel or similar tool. After the deposit is leveled, the depth of the deposit in several locations shall be measured (at least one location per 3 ft^2 of the sump) with a precision of 1 mm. The measured depths are averaged to determine the depth of the deposit prior to the test. After running water through the device for a period of approximately 1h, the sump is drained to about 1 cm above the deposit, the sediments in the sump are flattened and leveled, and the depth of the deposit is measured again at the same number of locations as in prior to the test, and the depth of deposit is estimated after the test. Effluent concentration of the washed out sediments will be estimated using the following equation:

$$C_{\text{eff}} = \frac{\Delta h \cdot A \cdot \gamma_w \cdot SG_{bw}}{Q\Delta t}$$

where Δh = difference between the deposit depth before and after the test; A = surface area of the sump; γ_w = specific weight of water; SG_{bw} = wet bulk specific gravity of the sediments; Q = flow rate; and Δt = duration of the test at the target discharge.

 - *Sampling Effluent Concentration.* Sampling of effluent concentration is not recommended due to pulsing of sediment during resuspension runs.

- The results of the sediment retention tests shall be plotted separately for the three PSDs tested. The plot will present effluent concentration versus discharge. Mean water temperature and the median particle size of the tested material shall also be reported on the plot. The washout rates of untested particle sizes with similar specific gravity can be approximated by interpolation.

1.5 FILTRATION TESTING

The purpose of this testing method guidance is to help decision makers better understand and document the performance of a full-scale filtration-based manufactured treatment device in a laboratory setting. To avoid potential error extrapolation within this protocol, testing a minimum of 0.35 m^2 or the entirety of filter modular unit surface area is suggested. Laboratory testing results can be used for design and scaling of field application, preparation of certain sediment reduction efficiency functions, and model simulations for specified rainfall and pollutant characteristics different from those tested.

1.5.1 Testing Objectives

The objectives of filtration testing are to

- Identify the hydraulic characteristics of the filter, including head, head loss, and headwater elevation at the specified maximum surface loading rate or hydraulic loading rate (flow rate per unit area) of a filtering device;
- Quantify the mass removal of a specified particle gradation at various surface loading rates (flow rate per unit area) for the filter device, while also quantifying the largest particle diameter(s) that may not be filtered; and

- Quantify the mass loading capacity of a filtering device of a specified particle gradation at the maximum surface loading rate (flow rate per unit area).

> **Health and Safety Advisory:** These filtration testing guidelines suggest using finely ground silica particulates to quantify performance of filtration devices. Fine silica particles (e.g., Sil-Co-Sil materials) contain crystalline silica-quartz, which can cause silicosis (an occupational lung disease) and lung cancer. Mitigation of dust generation and use of a respirator are required for safe application. For detailed information on the potential health effects of crystalline silica-quartz, refer to the Material Safety Data Sheet.

1.5.2 Testing Procedure

1.5.2.1 Hydraulic Characteristics and Medium Loss Testing

Hydraulic characteristics of the filtering medium typically change as they operate and ripen. Therefore, hydraulic characteristics should be identified on a fresh, unused filtering medium. Additionally, filtering media may have a tendency to release or lose part of the medium prior to it becoming stable within the system.

The following procedures should be used for testing medium loss:

1. Flush the system with fresh, clean water for as long as deemed necessary, working the surface hydraulic loading rate up to the designed maximum level. Record and report time and concentrations of the startup process.
2. Document head loss at the maximum surface loading rate ($L/s/m^2$ of the specified filtering mechanism) using clean water. Provide a plot of head loss vs. surface loading rate on a constant head basis.
3. Document headwater elevation throughout the test.
4. Sample effluent and test for filter medium loss, by testing SSC and PSD to determine if any filtering medium material has been lost during this clean water flush.

1.5.2.2 Removal Efficiency Testing

Manufactured filtration technologies are designed to provide coarse and fine solids removal with filter media. Filters may have the ability to capture finer particles than gravity separators. For influent concentrations of less than 100 mg/L, the testing facilities may pursue a sediment effluent concentration goal and a removal efficiency. Other laboratory testing protocols have used particle size gradations with a mean diameter in the 20 μm range. Suggested influent concentrations of tested sediment are 100 and 200 mg/L.

Removal efficiency should be tested at simulated storm events over surface loading rates of 25, 50, 75, 100, and 125% of the MDTR. Testing at lower surface loading rates may be deemed unnecessary, because filtration removal efficiency generally improves as surface loading rates decrease. Testing at greater than 125% of the MDTR is only necessary if a filter device is designed to be placed online.

The following are steps for removal efficiency testing:

- Elimination of background suspended particulates to less than 5 μm in size from any water used for this test procedure is recommended. This can be accomplished through avoidance of closed loop systems or water refiltration to ensure suspended particulates in the water loop do not interfere with the test or filtration device. Background SSC and PSD should be tested at least three times through the duration of the testing period and reported.
- It is recommended to use a fresh filtering medium or a backwashed filtering medium if repeatability can be demonstrated and documented. If backwashing has taken place, the following test should start at the same head as the preceding test, assuming the flow is within 10% of the stipulated surface loading rate. If the flow differential exceeds 10%, then either additional backwashing or fresh filtering medium is required.
- Removal efficiency at the maximum surface loading rate ($L/s/m^2$ of the specified filtering mechanism) using test sediment shall be reported as a percent removal (%) and as effluent concentration (mg/L).
- Sediment used for this laboratory test should be synthetic and silica in nature, having a narrow PSD, with a D_{50} of 38 μm or less, e.g., Sil-Co-Sil 106. The suggested silica test material that falls in the range classified as suspended and settleable sediment (<75 μm) is Sil-Co-Sil 52, which has a D_{50} of 12 μm. For improved feed accuracy and for safe use by mitigating small particulate dust generation, the suggested sediment feed should be slurried and then pumped as part of the influent feed.
- It is suggested to use influent concentrations between 100 and 200 mg/L.
- PSD testing is used to determine removal effectiveness of the finer particles. Influent sediment PSDs can be conducted by randomly collecting three samples from the full mixed sediment batch to be used, or all 12 influent samples can be analyzed. PSDs are to be tested on all 12 effluent samples. Laser diffraction or Coulter Counter can be used to quantify each distribution, to identify the largest diameter particle(s) not captured during each test condition, while also using the D_{50} of the PSD as an indicator of overall removal capability.
- Removal efficiency testing shall have at least 12 tests conducted total, with each test condition repeated at least twice per surface loading rate. For each test condition, five influent and effluent water samples will be taken and composited during the test and SSC and PSD analyses will be conducted. Removal efficiency testing shall be conducted starting from the lowest surface loading rate condition and increase progressively to the highest surface loading rate.

Note: To clearly identify sediment mass removal, SSC analysis is the preferred methodology. However, if desired, additional influent and effluent particulate related constituents could be analyzed, such as TSS [test method EPA 160.2/ (USEPA 1979)] or turbidity as long as a PSD such as Sil-Co-Sil 52 is used for testing.

- Headwater elevation should be recorded throughout each test condition and reported.
- The manufacturer's claim of maximum operating rate stated in terms of surface loading rate (L/s/m^2 of the specified filtering mechanism) at a given head shall be verified. Graphically plot removal efficiency versus surface loading rate and D_{50} removal efficiency versus surface loading rate:

$$RE = \frac{(Mi + BM + Me) \times 100}{Mi + BM}$$

where RE = mass removal efficiency; Mi = mass influent; BM = background mass; and Me = mass effluent.

1.5.2.3 Sediment Mass Loading Capacity Testing

In addition to testing removal efficiency, it is important to understand the sediment mass loading capacity of a filter system, as this is pertinent to the service life, end use application, and frequency of maintenance for a filter system. Sediment mass loading typically occurs over time within a treatment device. If a filter device is utilized downstream of a detention system, the controlling factor of long-term performance may not be removal efficiency, but mass load capacity. To determine overall capability and mass sediment capacity, all unit processes and potential cleaning mechanisms within a device or treatment train should be reasonably accounted for. For example, in addition to filtration, sedimentation remains a significant unit process in most filtration devices.

Testing sediment mass loading capacity should provide results indicating the sediment mass load removal capacity (kg/L/m^2) of the specified filtering mechanism for a given test sediment and head loss.

The following list outlines the steps for testing the sediment mass loading capacity:

- Repeat a clean water flush, and sample SSC and PSD to determine if medium loss has occurred.
- Elimination of background suspended particulates to less than 2 μm in size from any water used for this test procedure is recommended. This can be accomplished through avoidance of closed loop systems, or required water filtration to ensure suspended particulates in the water loop do not interfere with the test or filtration device. Background concentration and PSD should be tested at least three times through the duration of the testing period and reported.
- To evaluate the mass loading capacity of a filter device, simulating a series of "storm events" is recommended. This can be accomplished by mass loading

known sediment at an influent concentration range of 200 to 300 mg/L to accelerate the test. The characteristic of the sediment must be representative of what is typically found in the field, which may be regionally specific and agreed upon in the QAPP. The use of silica synthetic media is not representative of most field applications, which include bacteria, organic matter, and plate-*y* sediments.

- Each simulated "storm event" shall be conducted at the maximum surface loading rate held constant for 60-min intervals delivering a known cumulative amount of water volume and sediment mass load.
- After each simulated storm event, the system is permitted to automatically self clean and/or drain down as the device is engineered for up to a 24-h period to simulate the conclusion of a runoff event. When the self-cleaning and/or drain down period concludes, resume testing of the next 60-min storm event. The run-dry event cycle should be repeated until at least 75% of the maximum head of the device or a minimum of 60 h is reached, whichever happens first.
- To gain an indication of sediment mass capacity of solely the filtration device, measure headwater elevation continuously while conducting the simulated storm event. Graphically plot head loss versus cumulative sediment mass load, and extrapolate the plot to the 100% head loss point. For constant head systems, plot cumulative load versus flow rate.
- To gain an indication of sediment removal efficiency over time (improvement or reduction) of solely the filtration device, measure SSC and effluent concentration over a minimum of eight equal intervals by collecting three effluent samples up to the point where 75% of the maximum head is achieved. Graphically plot removal efficiency versus head loss and removal efficiency versus cumulative sediment mass load.
  Note: It is also suggested that PSD be tested from each sample in addition to SSC, or from an overall composite sample to determine if medium loss has occurred.
- Report cumulative mass sediment captured per filter element (e.g., cartridge) when 75% of the maximum head of the device is reached, or when 60 h are reached, whichever happens first. Report in a format that shows pounds per square feet (lb/ft^2) surface area and per element.
- Provide an accounting of how the mass is distributed in the "system." For example, how much is in the tank versus in the filter. Characterize the central cause for occlusion (e.g., surface occlusion, media occlusion). Sections of the media should be observed and possibly measured to characterize the distribution of the sediment within the filter.

1.5.3 Advisory on Testing of Bypasses

Most filtration devices are placed off-line or have internal bypasses to protect the filtering medium and extend the performance capability. To determine a filtration device's overall performance if an in-line device has an internal bypass, the device shall be subject to additional performance monitoring of hydraulic characteristics to at least 125% of the surface loading rate and a **scour laboratory test evaluation** following the same protocol requirements as a hydrodynamic sedimentation device conducted at the 100 and 125% maximum surface loading rate [L/s/m^2 (gpm/ft^2)].

1.6 QUALITY ASSURANCE AND CONTROL

Laboratory data shall be collected under a quality assurance/quality control (QA/QC) plan. The QA/QC plan shall include the following as a minimum:

1. Project Description
2. Project Organization and Responsibility
3. Data Quality Objectives
   a. A list of parameters being tested, the minimum requirements in terms of sample numbers, and other numeric criteria to qualify the data.
4. Method Quality Objectives
   a. A description of the methods used to ensure that collection of data has sufficient quality assurance and control and verifiable independence.
5. Project Test Methods
   a. Methods to adjust for expected background concentrations of material in inflow test water.
   b. Calibration of the system and equipment used to dose sediment, including flow rates to ensure that influent concentrations are maintained within test parameters and methods for accurately measuring the mass of sediment added to the influent pipe.
   c. Equipment cleaning and blanks.
   d. Verification of sampling techniques and sample preservation methods.
   e. Chain of custody.
6. Laboratory Procedures
   a. Constituents for analysis.
   b. Laboratory performance standards.
   c. Analysis method references.
   d. Data reporting requirements.
   e. Data validation procedures, such as including blanks and running duplicates.
   f. Corrective actions.

1.7 REPORTING TEST RESULTS

It is recommended that the laboratory report contain at a minimum the following information.

Executive Summary

Background

Name of laboratory, principal investigator, and subcontractors.
Qualification statements for laboratory and data analysts.
Laboratory equipment. List including

- Name, model, and dimensions (depth and height) of the device tested;
- Pumps, compressors, mixers, valves, and flow and water quality sampling equipment;
- Storage tanks;
- Standpipe and plunge pool; and
- Filtration equipment.

Settling chamber diameter (L_1) and depth (L_2) measurements.
Inlet and outlet pipe dimensions.
Modifications. Any made to the device to enhance transportation or test feasibility and explanation of why these modifications are not expected to affect the laboratory results.
Process flow diagram. Showing test device, piping, water source, pump, storage tanks, sediment injection system, sampling locations (if sampling technique has been adopted), and flow meter.

Efficiency Testing and Results

The following shall be reported for each device tested:
Date, flow rate, and elapsed time for the test. Flow rate needs to be reported in terms of both per individual unit (e.g., gpm/unit) and in terms of flux or surface/hydraulic loading rate (e.g., gpm/ft^2)
Tabular results of test parameters. Where particle size data are required, it shall be reported for each of the particle sizes tested (in μm).

- Test Sediment Used: Total dry mass of test sediment placed in the sediment hopper, total dry mass remaining in the hopper, and total dry mass (calculated by difference) of test sediment discharged from the hopper during the test. If a mix of several discrete particle sizes is introduced, the entire mass introduced into the hopper should be fed into the system.
- Test Sediment Retained: Total dry mass of test sediment removed from the settling chamber. Component mass by particle size of sediment removed from the settling chamber.
- Effluent Sampling Results: If a verified sampling method is used for measuring concentrations, then for each discrete effluent sample, the SSC, the

component mass, and concentration by particle size class (can be done through a PSD analysis).

Performance efficiency. Dry mass retained. Tabular data for each test flow showing the calculated percent reduction based on mass entering the device and mass retained. Calculations shall be performed by total mass and by particle size class.

- Percent reduction shall be based on a comparison of mass of sediment introduced to the sediment chamber and mass of sediment retained in the sedimentation chamber, where

$$\%\text{Reduction} = (\text{mass retained}/\text{mass fed}) \times 100$$

  If a verified sampling method has been utilized, then tabular data shall be used for each test flow showing the calculated percent reduction in mass of test sediment based on inlet and outlet concentrations reported in the "Tabular results of test parameters" section of the laboratory report. Calculations shall be by total mass of the average equivalent diameter (D_{50}) of each particle size class.

$$\%\text{Reduction} = (\text{Conc}_{\text{in}} - \text{Conc}_{\text{out}})/\text{Conc}_{\text{in}} \times 100$$

- The tabular analysis is recommended to be presented in the format of Table 1.2.
  - The tabular data set shall also be presented in graphical form. A separate graph for the D_{50} of each particle size class shall be presented that shows the percent reduction (y) as a function of flow rate (x) for the D_{50} of the particle size class and the range of temperature tested. See Fig. 1.1 as an example.
  - Manufacturers are encouraged to use an approach that could provide a predictive formula, i.e., a performance function, for the device at all flow conditions, particle sizes, particle densities, and water temperatures. For HDSs, a method has been developed by Wilson et al. (2009) that

Table 1.2. Format for test results

	% Reduction by median particle size					
Flow Rate	D_1 *(microns)*	*Temperature (°C)*	D_2 *(microns)*	*Temperature (°C)*	D_3 *(microns)*	*Temperature (°C)*
0.25 × MDTR[a]						
0.50 × MDTR						
0.75 × MDTR						
1.00 × MDTR						
1.25 × MDTR						

[a]MDTR = maximum design treatment rate as determined by the manufacturer

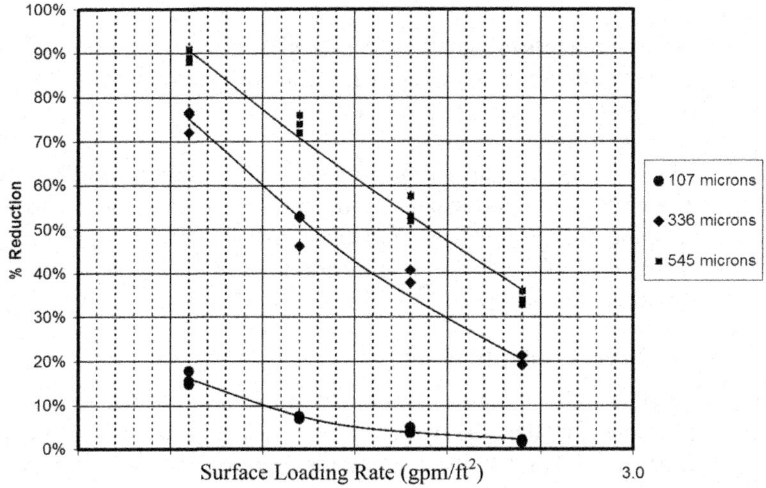

Fig. 1.1. Example of a graphical presentation of removal efficiency (percent reduction) versus discharge for the D_{50} of three particle size classes

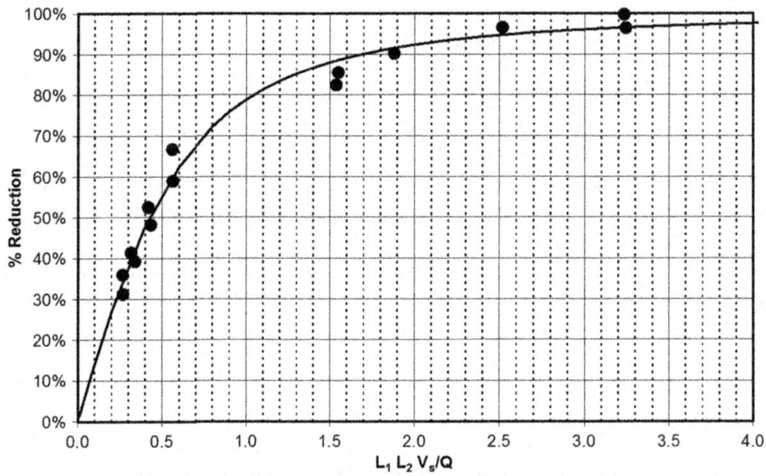

Fig. 1.2. Example of a graphical presentation of removal efficiency (percent reduction) as a function of the dimensionless number proposed by Wilson et al. (2009)

incorporates all these factors into a single dimensionless number. By fitting a function to the data, removal efficiency can be explained as a function of the dimensionless number as follows:

$$\%\text{Reduction} = \text{Function}\left[(L_1 * L_2 * V_s)/Q\right]$$

Where
L_1 = Device characteristic length 1, e.g. height;
L_2 = Device characteristic length 2, e.g. depth of the sump;
V_s = Particle size settling velocity; and
Q = Discharge through the device.
Different equations are found in the literature to estimate the settling velocity (e.g., Cheng 1997). If a performance function is developed, the equation used for developing the performance function shall be used for predicting the removal efficiency of the device in other conditions. Fig. 1.2 shows a graphical presentation of this approach.

Signatures for Report Submittal

The report shall be signed by the laboratory director or his/her designee, the person responsible for data analysis and reporting, and, if applicable, the independent observer. The technical standards used for testing should be listed. The signers shall attest that the laboratory testing and data analysis have been conducted in accordance with the requirements of the pertinent technical standards.

References

ASTM. (2007). "Standard test method for determining sediment concentration in water samples." *ASTM D3977-97*, West Conshohocken, PA.

ASTM. (2011). "Standard test method for measurement of hydraulic characteristics of hydrodynamic stormwater separators and underground settling devices." *ASTM C1745/C1745M–11*, West Conshohocken, PA.

ASTM. (2012). "Standard test method for measurement of suspended sediment removal efficiency of hydrodynamic stormwater separators and underground settling devices." *ASTM C1746/C1746M-12*, West Conshohocken, PA.

ASTM. (2016). "Standard test method for measurement of hydraulic characteristics of stormwater filtration elements." *ASTM C1814/C1814M-16*, West Conshohocken, PA.

Cheng, N. S. (1997). "A simplified settling velocity formula for sediment particle." *J. Hydraul. Eng.*, **123**(2), 149–152.

Lenhart, J. H, Ogintz, J., and de Ridder, S. (2009). "Design optimization of hydrodynamic separators." *Proc., 2009 ASCE-EWRI Conf.*, ASCE, Reston, VA.

Roesner, L. A., Pruden, A., and Kidner, E. M. (2007). *Improved protocol for classification and analysis of stormwater-borne solids*, Water Environment Research Foundation, Alexandria, VA.

USEPA (U.S. Environmental Protection Agency). (1979). "Methods for chemical analysis for water and wastes." *EPA-600/4-79-020*, Office of Research and Development, Cincinnati.

Wilson, M., Mohseni, O., Gulliver, J., Hozalski, R., and Stefan, H. (2009). "Assessment of hydrodynamic separators for stormwater treatment." *J. Hydraul. Eng.*, **135**(5), 383–392.

CHAPTER 2

Scaling Relations

Physical Scaling Subcommittee

Scaling criteria helps to develop performance specifications for various sizes of manufactured treatment devices (MTDs) and to utilize model-scale tests to develop performance specifications. However, an industry-wide, uniform methodology to scale and size these devices does not currently exist. This section will review the need for scaling criteria in MTDs and propose scaling criteria for head loss and flow, particle separation, scour, and filtration.

2.1 NEED FOR SCALING RELATIONS

Because of interaction with a complex flow field, the need for scaling criteria is probably greatest in hydrodynamic separators (HDSs) and other underground stormwater control measures (SCMs). The vertical velocity relative to water causes separation while the turbulence of the flow field tends to mix the medium that is being separated back into the water. Three issues need to be addressed: (1) settling of heavier-than-water particles out of the flow, (2) rise of lighter-than-water compounds such as oils and trash out of the flow field, and (3) the tendency to entrain both in high flow after separation has been accomplished. Experimental tests are possible on model-scale or full-scale devices, but these tests will need to be scaled to another size of device to meet demand. In addition, if an acceptable scaling procedure can be developed and verified, it will be possible to perform smaller-scale model studies on each device, which can be scaled to full device size with a known accuracy.

A filter medium can be tested through column studies. However, the parameters that need to be controlled (e.g., grain size versus column diameter) and scale-up to full-scale filters needs to be documented. In addition, many filters are partially self-cleaning, utilizing turbulence in the flow field to remove caked material from the filter. The scaling relations described above may be sufficiently general to apply to this case as well. Tests would still be required on the devices to demonstrate and measure the self-cleaning properties. To apply the results of such tests to multiple device sizes would require scaling criteria.

2.2 DIMENSIONLESS NUMBERS AND SCALE EFFECTS

One goal of scaling is to have one dimensionless number that represents the important processes of the device. This is not generally possible at all scales, so criteria are developed on the appropriate model scale. These criteria include a critical Reynolds number (R) value that must be exceeded to ensure that turbulent flow occurs,

$$\mathsf{R} = \frac{Q}{\nu L_S} \geq \mathsf{R}_C \approx 5{,}000 \tag{2-1}$$

and a critical Weber number (W) that must be exceeded to ensure that surface tension does not alter the flow field (ASCE 2000),

$$\mathsf{W} = \frac{\rho Q^2}{\sigma L_S^3} \geq \mathsf{W}_C \approx 100 \tag{2-2}$$

where Q = discharge through the device; ν = kinematic viscosity; ρ = density of the liquid; σ = liquid surface tension; and L_s = smaller of the important length scales in the flow field, such as the diameter of the device, annular clearance, or depth of water. In addition, it may be necessary to enhance the wall roughness in a physically-scaled model to result in a similar relationship of wall shear stress to flow velocity (similar friction factor or drag coefficient) as in the full scale.

Fenner and Tyack (1997) reviewed the scaling criteria for removal efficiency and head loss of HDSs and, in addition to geometric similarity, advocated for the following dimensionless similarity criteria.

2.2.1 Removal Efficiency

Hazen number $$\mathbf{Ha} = \frac{AV_S}{Q} \tag{2-3}$$

Froude number $$\mathsf{F} = \frac{Q}{\sqrt{gd^5}} \tag{2-4}$$

2.2.2 Pressure Loss

Euler number $$\mathbf{Eu} = \frac{\Delta P d^4}{\rho Q^2} \tag{2-5}$$

Friction factor $$\mathbf{f} = \frac{u_*}{V} \tag{2-6}$$

where A = plan area of the separator; V_s = settling velocity of the particles; V = mean flow velocity; ΔP = pressure drop; d = length dimension, and

$u_* = (\tau_{wall}/\rho)^{1/2}$ = shear velocity at the walls. The Hazen number, Ha (Bloodgood et al. 1956), is designed for similarity of settling in a quiescent flow field (Hazen 1904); Froude number (F) similarity is required for fluid flow driven by gravity; Euler number similarity will appropriately scale up pressure loss; and friction factor similarity is required for the influence of turbulence in the flow field. Friction factor similarity would also be important for removal efficiency, because of the impact of turbulence on particle settling.

Based upon previous experience with settling of suspended sediments in lakes (Dhamotharan et al. 1981), the Péclet number ($\mathsf{P} = V_s L/D_t$) with a dimensionless time is an appropriate dimensionless scaling parameter to explain sediment deposition rates. In the definition of the Péclet number (P), L is vertical settling scale, and D_t is a vertical eddy diffusion coefficient. P is defined as the ratio of convection to diffusion; in this case the convective, settling process is opposed by eddy diffusion in the system tending to keep solids in suspension. Similar to the approach taken for lakes and reservoirs, Wilson et al. (2007, 2009) scaled convection with the settling velocity (V_s), used the settling depth (h) as a vertical length scale L, and a flow velocity times a length scale for the vertical diffusion coefficient (D_t). When the diameter is the shortest dimension of the flow, $D_t \sim Qd/A$. In most HDSs, flow enters the chamber from above and turbulence is stronger where plunging occurs. Therefore, if A is taken as the horizontal projection of the chamber (the cross-sectional area of mixing), which is proportional to d^2, then $D_t \sim Qd/d^2 = Q/d$. The Péclet number for the stormwater treatment device therefore becomes

$$\mathsf{P} = \frac{V_s hd}{Q} \tag{2-7}$$

If the smallest dimension of the device is its height or settling distance, as is the case for detention ponds, then inflow is relatively horizontal and A is best characterized as the cross section of the pond, hd. Therefore $D_t \sim Qh/dh = Q/d$ and $\mathsf{P} = V_s hd/Q$. In both cases, one obtains the same equation for the Péclet number. The resulting Péclet number resembles the Hazen number, $V_s d^2/Q$, which is the ratio of settling velocity to overflow rate and was not developed to scale eddy diffusion. As long as geometric scaling is followed, it makes no difference whether one uses the Hazen number or the Péclet number. It might make a difference, however, in determining a dimensionless number to plot experimental data.

2.3 SCALING OF REMOVAL EFFICIENCY FOR HYDRODYNAMIC SEPARATORS

2.3.1 Removal Efficiency

Fenner and Tyack (1997) investigated the performance of 1.6- and 0.3-m diameter HDSs using plastic beads of specific gravity 1.04. They found that Froude scaling was most accurate at low removal efficiencies and Hazen scaling was more successful at higher removal efficiencies. They proposed a hybrid scaling formula, which incorporates the impact of each dimensionless number:

$$\frac{Q_A}{Q_B} = \eta_B L_r^2 + (1 - \eta_B) L_r^{2.5} \left(1 + \frac{d_p}{3 \text{ mm}}\right) \tag{2-8}$$

where the subscripts A and B indicate two different scales, or prototype and model, respectively; η = fraction of removal; d_p = mean particle diameter in mm; and L_r = length ratio (Scale A/Scale B). To be used in practice, Eq. (2-8) should be adjusted for the specific gravity of the particles, which typically will differ from those used in Fenner and Tyack's experiments, or arranged to use the settling velocity of the particles. The latter was performed on Fenner and Tyack's data to result in

$$\frac{Q_A}{Q_B} = \eta_B L_r^2 + (1 - \eta_B) L_r^{2.5} \left(1 + \frac{V_S}{25 \text{ mm/s}}\right) \tag{2-9}$$

where V_S = settling velocity of the particles in mm/s.

A critique of Eq. (2-8) is that it mixes flow scaling with particle settling into one equation. The result is that Eqs. (2-8) and (2-9) will be applicable only to the particles that Fenner and Tyack tested. This is not necessary, and it is possible to separately scale settling velocity and flow. One proposal for scaling is outlined in the following sections.

2.3.2 Geometric Scaling

Geometric scaling is crucial to similarity. Simply stated, all important geometries of the MTD must be scaled proportionately, such that L_r is the same value for every dimension in the MTD.

2.4 FLOW AND HEAD LOSS IN HYDRODYNAMIC SEPARATORS

Flow in most HDSs is controlled by gravitational forces. The field of physical hydraulic models has been performing this type of scaling for many years. The following criteria are commonly used to ensure valid results:

1. The model should be sufficiently large that surface tension (Weber number effects) does not alter performance and scaling of the model, or $W > 100$.
2. Reynolds numbers in the model should be sufficiently large such that the flow is well into the turbulent range, or $R > 5{,}000$.
3. The friction factor in the model should be close to the prototype value. This often means adding some roughness to the model. This will make the ratio of turbulence to flow velocity similar in the model and prototype.

4. Froude number scaling is now possible,

$$\mathsf{F}_A = \mathsf{F}_B \tag{2-10}$$

Then,

$$\frac{Q_A}{Q_B} = \left(\frac{L_A}{L_B}\right)^{5/2} \tag{2-11}$$

2.5 SETTLING AND RISE IN HYDRODYNAMIC SEPARATORS

Scaling for settling with both retention (Hazen number) and turbulent mixing (Péclet number) result in a similar scaling parameter for settling of solids in HDSs. Froude number scaling, Eq. (2-11), should also be followed. The final result is

$$\frac{Q_A}{Q_B} = \frac{V_{sA}}{V_{sB}}\left(\frac{L_A}{L_B}\right)^2 \tag{2-12}$$

It is possible to satisfy both Eqs. (2-11) and (2-12) because of the addition of settling velocity to Eq. (2-12). In Eq. (2-12), that both mixing and retention result in the same scaling parameters is encouraging. In addition, rise velocity for floatables should follow the same scaling parameters as settling.

Several relationships for settling velocity have been developed (ASCE 2008, pp. 41–43). Ferguson and Church (2004) developed a relatively simple relation:

$$V_s = \frac{gRd_p^2}{18\nu + (0.7a5CgRd_p^3)^{1/2}} \tag{2-13}$$

where R = specific gravity in water (1.65 for silica sand) and C = constant equal to 0.4 for spheres and 1 for typical sand grains. Equation (2-13) approaches Stokes' law at small particle diameters and a constant drag coefficient for large particle diameters. This is the proper performance at the limits and is the reason that this equation is preferred. There are differences in the various relationships, so one must be careful to be consistent in the use of one relation.

2.6 SCOUR IN HYDRODYNAMIC SEPARATORS

Collected solids in HDSs, if not cleaned, will eventually reach the height where scour of the collected solids will occur. In addition, some HDSs are exposed to higher flows, beyond the design flow, and are therefore subject to potential scour of collected solids. Scour is physically modeled by equating (dimensionless) Shields stress, the presumed bottom stress where sediment begins to move off of the bottom, divided by the density and equivalent particle diameter of the sediment. Shields stress for uniform flow conditions is given by

$$\tau^* = \frac{u_*^2}{gRd_p} \tag{2-14}$$

where

$$u_* = \sqrt{\frac{f}{8V}} \tag{2-15}$$

indicates a bottom shear velocity.

Equating Shields stress for two different scales results in

$$\frac{Q_A}{Q_B} = \sqrt{\frac{f_B d_{pA} R_A}{f_A d_{pB} R_B}} \left(\frac{L_A}{L_B}\right)^2 \tag{2-16}$$

As with settling, F scaling [Eq. (2-11)] should also be followed. The process of scour in an HDS is a combination of lifting sediment off of the bed and settling of that sediment back to the bed. Thus, scour similarity needs to simultaneously satisfy Eqs. (2-12) and (2-16). To get similarity relations in Eqs. (2-12) and (2-16) to be satisfied simultaneously, relatively large sediment is necessary, such that Eq. (2-13) becomes

$$V_s = \sqrt{\frac{gRd_p}{0.75C}} \tag{2-17}$$

This is often difficult to achieve simultaneously with Froude number scaling. All sediment that is washed out of an MTD would go through a two-part process: First, it would need to be scoured from the sediment bed. Second, it would need to be suspended toward the outlet of the MTD. It has not been determined whether bed load transport, which would scale with Shields stress, or suspended load transport, which would scale with the Hazen number, is dominant in sediment wash out. If it is the former or both, the problems of matching Eqs. (2-12) and (2-16) exist for proper scaling of sediment scour and wash out.

2.7 REMOVAL EFFICIENCY FOR FILTERS

Scaling of filter modules tends to be linear, that is, double the flow, then double the number of filters. Scale effects tend to relate to flow rates and media size, hence this discussion focuses on those two parameters. The particle trapping of a filter is a local phenomenon, dependent upon the characteristics of the media and the particle size distribution of the suspended solids. This can be relatively well simulated with a bench-top column test, where the media filtration efficiency is tested with the particle size distribution of interest. The scaling parameters of most interest are the media Reynolds number,

$$\mathsf{R}_m = \frac{Q}{\varepsilon^{1/3} A_{cs}^{1/2} V} \tag{2-18}$$

and the suspended solid-media size ratio,

$$L_{\text{pm}} = \frac{d_p}{d_{\text{med}}} \tag{2-19}$$

where ε = porosity of the filter media; d_{med} = equivalent spherical diameter of the media; and A_{cs} = cross-sectional area of filter. Then, scaling of discharge and media would follow the relations

$$\frac{Q_A}{Q_B} = \left(\frac{\varepsilon_A}{\varepsilon_B}\right)^{1/3} \frac{L_{\text{pm}A}}{L_{\text{pm}B}} \frac{V_A}{V_B} \tag{2-20}$$

and

$$\frac{d_{\text{med},A}}{d_{\text{med},B}} = \frac{d_{p,A}}{d_{p,B}} \tag{2-21}$$

The scaling of one particle size distribution to another and scaling with filter flow rate are of interest because they would reduce the required filtration tests. In addition, filters are often placed in the turbulent flow field of the device, resulting in differential pressure drop across the filter that varies spatially.

In the application of filters, one also needs to consider the fouling that will occur at the filter surface. Fouling will increase the frequency of cleaning and could reduce the filter life. The cleaning of particle cake that has accumulated on the surface of the filter, however, cannot be simulated with column tests because cleaning is often provided by turbulence generated in the chamber of the filter device. This would require a scaling analysis similar to high flow scour, because both are based upon the resuspension of collected material. It should be noted that some laboratory testing at the upper and lower bounds is needed to support these theoretical predictions.

2.8 SUMMARY OF PROPOSED SCALING RELATIONS

The proposed scaling relations include the following:

- Scaling laws have been proposed for head loss, settling, scour, and filtration. These laws need to be tested with combined model and prototype studies of manufactured SCMs. *Until these proposed scaling laws can be properly tested and verified, scale model testing should only be used as a diagnostic technique to observe particle movement.*

- The scaling relation of Fenner and Tyack is not recommended, because it combines settling and head loss similarity and is applicable only to the type of spheres used in their experiments.
- Froude number scaling, with approximately equal friction factors, should be used in all model studies.
- Settling can be scaled with Péclet number or Hazen number. They are equivalent as long as geometric scaling is used and encompasses the important processes of settling.
- Scaling for scour should follow Shields stress and Hazen (Péclet) number scaling. This will be difficult, but is possible if the sediment is sufficiently large, such as with plastic beads.
- The scaling of filtration units is dependent upon the size of the particles relative to the size of the media and the pressure distribution in the filtration chamber. Froude scaling with equal friction factors should be sufficient for pressure distribution. In addition, cleaning of a filter unit is similar to scour, such that the scour scaling parameters should also be followed to simulate cleaning of the filter fouling. *While bench tests are acceptable to test the filter media, the operation of the filter as a whole deserves a full-scale prototype test, because of the complexities of pressure distribution and fouling.* Model scaling relations would help in determining the performance of manufactured filtration units.

2.9 NOTATION

A	plan area of the separator
A_{cs}	cross-sectional area of filter
C	constant equal to 0.4 for spheres and 1 for typical sand grains
d_{med}	equivalent spherical diameter of the media
d_p	mean particle diameter
D_t	vertical eddy diffusion coefficient
Eu	Euler number $= \Delta P d^4/(\rho Q^2)$
f	Friction factor $= u_*/V$
F	Froude number $= Q/(gd^5)^{1/2}$
g	acceleration of gravity
Ha	Hazen number $= V_s A/Q$
L	vertical settling scale
L_r	length ratio $= \text{scale}_A/\text{scale}_B$
L_{pm}	suspended solid-media size ratio $= d_p/d_{med}$

L_s	smaller of the important length scales in the flow field
P	Péclet number $= V_s L/D_t => V_s hd/Q$ for a settling MTD
Q	discharge through the device
R	specific gravity in water (1.65 for silica sand)
R	Reynolds number $= Q/\nu d$
R_m	media Reynolds number $= Q/(\varepsilon^{1/3} A^{1/2} \nu)$
u^*	$(\tau_{wall}/\rho)^{1/2}$, the shear velocity at the walls
V	cross-sectional mean flow velocity
V_s	settling velocity of the particles
W	Weber number $= \rho Q^2/(\sigma d^3)$
ΔP	pressure drop
ε	porosity of the filter media
η	fraction removal
ν	kinematic viscosity of the liquid
ρ	density of the liquid
σ	liquid surface tension
τ^*	Shields stress

References

ASCE. (2000). *Hydraulic models: Concepts and practice*, R. Ettema, ed., Reston, VA.

ASCE. (2008). *Sedimentation engineering: Processes, measurements, modeling and practice*, M. H. Garcia, ed., Reston, VA.

Bloodgood, D. E., Boegly, W. J., and Smith, C. E. (1956). "Sedimentation studies." *Proc. Am. Soc. Civil Eng.*, **82**(SA5), 90–93.

Dhamotharan, S., Gulliver, J. S., and Stefan, H. G. (1981). "Unsteady one-dimensional settling of suspended sediment." *Water Resour. Res.*, **17**(4), 1125–1132.

Fenner, R. A., and Tyack, J. N. (1997). "Scaling laws for hydrodynamic separators." *J. Environ. Eng.*, **123**(10), 1019–1026.

Ferguson, R. I., and Church, M. (2004). "A simple universal equation for grain settling velocity." *J. Sediment. Res.*, **74**(6), 933–937.

Hazen, A. (1904). "On sedimentation." *Trans. ASCE*, **53**, 45–71.

Wilson, M. A., Mohseni, O., Gulliver, J. S., Hozalski, R. M., and Stefan, H. G. (2009). "Assessment of hydrodynamic separators for stormwater treatment." *J. Hydraul. Eng.*, **135**(5), 383–392.

Wilson, M., Gulliver, J. S., Mohseni, O., and Hozalski, R. M. (2007). "Performance assessment of underground stormwater treatment devices." *St Anthony Falls Laboratory Project Rep. No. 494*, Univ. of Minnesota, Minneapolis, MN.

CHAPTER 3

Field Testing

Field Testing Subcommittee

Field testing of manufactured treatment devices (MTDs), an industry-specific class of unit operations and processes, for management of particulate matter (PM) requires planning and implementation through all steps of the verification process. The basic steps of field testing of MTDs for verification can be summarized as follows: (1) watershed selection, (2) testing, (3) analytical protocols, (4) data evaluation, (5) quality assurance and control measures, and (6) data reporting. Field monitoring should provide verification of analytical or numerical models for MTDs that have been developed through scaled or full-scale controlled physical model testing. The goals of the field verification must be designed into a certification process. Ultimately, the process leads to a defensible set of results for an MTD across a wide range of event-based loadings. In this context, the number of events for certification should be no less than 15 with a minimum time between sampling events of 6 h.

These Guidelines focus on field testing of MTDs for management of PM from rainfall runoff and snowmelt. Toward this goal, the entire gradation of PM requires characterization for an MTD field evaluation from gross solids such as biogenic material to suspended PM. The suspended fraction is of particular interest because of its mobility and acute bioavailability. However, the gross solids and sediment fraction are important for regular maintenance because these coarse fractions fill many MTDs volumetrically and are most labile and leachable.

A fundamental, transferable, and defensible methodology is discussed in these Guidelines. This methodology includes particle size distributions (PSDs) combined with gravimetric index tests for PM and turbidity. The suggested gravimetric index tests include total suspended solids (TSS) and suspended sediment concentration (SSC), the suspended, settleable, and sediment fractions. Despite controversy, TSS remains in circulation because of ubiquitous usage and regulatory significance and as an index of treatability. SSC provides a gravimetric analysis of the entire sample with limited bias and improved reliability and repeatability. Testing must elicit the role of variability and collect sufficient representative data within an event and across events so that the variability of the MTD-load response can be quantified.

The field verification process must provide the rigor required for the defensible and transferable examination of MTD behavior with respect to PM and flow regardless of the region and watershed where testing is conducted. This complexity extends to watershed characteristics, rainfall-runoff event characteristics, operation and maintenance requirements, system sizing tied to hydrologic and PM loadings, testing methodologies including mass balances, and the nonstationary behavior of the MTD. The intent of this chapter is to outline a verification protocol for PM that is transferable and defensible for a broad set of loading conditions so that there will be no need of state-to-state field verification of MTDs.

3.1 COMPONENTS OF FIELD VERIFICATION

3.1.1 Watershed Site Selection

To monitor an MTD in the field, it is critical to choose a location representative of the expected operating conditions of the MTD to be evaluated and the region of the certifying entity. Ideally, there will be enough background information to allow the researcher to predict the average flow and water quality variables to a degree that provides accurate results. Installing an MTD into the ground for the specific purpose of a monitoring program allows control of some of the physical variables that inhibit testing; but unless there is a dedicated testing site for all MTDs, there will be variable land uses and pollutant loadings from real-world installations.

When selecting a location for the MTD, it is important to determine the drainage area, average rainfall intensity, peak flow, and other precipitation (including snow and ice) and physical characteristics of the site that will affect the performance of the MTD. Optimally, the site will have on average at least eight storm events per year (for a minimum of 15 storm events in two years) to have a sufficient number of samples for verification. Note that the required number of storm events might need to be assessed and modified in dry climates with very low annual precipitation. The site selection criteria should also include the following conditions to allow for control of pollutant loadings and setup and maintenance of sampling equipment:

1. Minimize equipment requirements by using an MTD with one influent pipe and one effluent pipe, each pipe with an autosampler. Use only one flow meter if the device is *run of river*. However, if storage or flow attenuation is part of the MTD function, two flow meters may be required to operate the autosamplers independently. Additional pipes will require additional autosamplers and flow meters. More equipment leads to more malfunctions and lost storm sampling opportunities.
2. There should be no base flows through the pipes or backwater or submersion from downstream water bodies.
3. There should be no bypass flows for storms less than design, and overflows should not occur. If overflows do occur, the overflow needs to be part of the effluent sample.

4. MTDs located in roadways should be avoided. Technician vehicles will need to be parked next to the site for sampling and equipment maintenance. Access dictates a location outside of the pavement for safety reasons and to avoid lane closures.
5. For rain gauges and solar panels to operate accurately there can be no tree coverage over the gauge or the looking angle of the solar panel. Rain gauges must be within the MTD's watershed.
6. Theft-proof enclosures should be used to house autosamplers, batteries, and solar panels. Adjacent property owners should be canvassed to ensure their cooperation with technicians accessing equipment at any hour.
7. Testing for gross solids requires selecting a watershed with high coverage of tree canopy.
8. The interior of the MTDs should have sufficient clearance and access to enable a technician to install equipment and take samples. The sites should be within reasonable driving distance for technicians who will perform inspections, provide equipment calibration and repairs, and retrieve samples and reset the autosampler. Automated sampling equipment capable of relaying alerts via modem or the internet is preferred to minimize site visits for stormwater events that do not trip the autosampler (due to low intensity or short duration for example).
9. Roadways in the watershed should have curb and gutters. There should be no other upstream stormwater control measures in the drainage basin, including roadside swales that will filter pollutants, especially gross solids and sediments, before they enter the MTD.
10. It is recommended that the maximum design treatment rate for the device being tested falls between 75% and 125% of the expected peak runoff rate expected for the site. The protocol needs to dictate that verification and certification of the technology are only given to the highest flow measured up to design. So for example, an MTD designed to treat 1 cfs is only monitored at 0.5 cfs; then it is verified at 0.5 cfs, not 1.0 cfs.

3.1.2 Quality Assurance Project Plan

Once the site has been selected, a quality assurance project plan (QAPP) needs to be prepared for review and approval by the certifying entity. A QAPP is required for each test site. The QAPP must specify the verification procedures to be followed to ensure the validity of the test results and conclusions. A QAPP addresses the following basic elements:

- Who will use the data and how will it be used?
- What are the goals and objectives of the project?
- What decisions will be made from the information obtained?
- How, when, and where will project information be acquired or generated?

- What potential problems may arise? What actions can be taken to mitigate their impact on the project? What are conditions triggering abandonment of the study?
- What type, quantity, and quality of data are specified?
- How defensible are the data to support the decision to be made?
- How will the data be analyzed, assessed, and reported?

A person with qualifications in chemical analytical methods should develop the QAPP in consultation with the analytical laboratory. The QAPP should include a description of the field sampling and data validation procedures. Standardized and published methods and procedures should be used to collect MTD data. The QAPP should be kept current and when a substantive change is warranted, the QAPP is to be revised and resubmitted for approval. The four basic elements of a QAPP are

1. Project management;
2. Data generation and acquisition;
3. Assessment and oversight; and
4. Data validation, usability, and transferability of the data.

EPA Requirements for QA Project Plans (QA/R-5) (available at http://www.epa.gov/quality/qa_docs.html) can be used as a basis of elements to be included in the QAPP. A sample QAPP for an MTD verification study is included in the appendix as an example.

3.1.2.1 Site Information

The QAPP shall contain at a minimum the following information about the proposed site:

- Drainage area.
- Breakdown of cover conditions (areas by type and condition of cover).
- Topographic survey.
- Drainage system drawing, including pipe sewers, appurtenances, and surface drainage components.
- Distribution of rainfall depth on a monthly basis.
- Temperature variability on a monthly basis.
- Soil type.
- Source of irrigation water (if any) for drainage area. Special emphasis should be placed on irrigation with reclaimed wastewater and hours and rates of application of irrigation water. While this may be of interest and could affect other variables besides sediment, there is no practical way to quantify the impacts of nutrient-laden irrigation runoff.

3.1.2.2 Device Information

The QAPP shall contain at a minimum the following information on the device to be tested:

- CAD representation of MTD system with detailed dimensions as proposed for the site.
- Schematic graphical representation of the MTD system.
- Manufacturer's drawings/specifications.
- A tabular summary of previous controlled laboratory testing results with respect to flow rate, PM, and PSDs.

3.1.2.3 Monitoring Information

The QAPP shall contain at a minimum the following information on the proposed monitoring setup:

- Aerial view of site watershed with drainage area delineated, scale and north arrow shown.
- Plan and profile of proposed MTD installation with inflow and outflow geometries (pipe diameter, slope, material, elevations, etc.).
- Tabular specifications on flow measuring devices and specification sheet for each device.
- Specifications on data acquisition systems and flow chart logic for sampling.
- Sampler specifications and locations and sampler specification sheet.
- Sample bottle specifications.
- Parameters to determine which storms are suitable for sampling, including minimum depth, minimum intensity, etc.

3.1.2.4 Sampling and Analysis Plan Information

The QAPP shall contain at a minimum the following information on the proposed sampling and analysis plan:

- Standard operating procedure (SOP) and method reference for determining TSS.
- SOP and method reference for determining SSC.
- SOP and method reference for determining PSD.
- SOP and method reference for determining PM fractions.
- SOP and method reference for determining turbidity.
- Equipment and sampling decontamination and preparation for an event.
- Sample preservation (if any) and sample holding time (< 24 h) before analysis.

- Sample volume and replicates.
- QC samples (spikes of known PSD or mass, blanks, splits, and field and lab replicates).
- QA on sampling equipment (calibration of automatic samplers and calibration of flow measuring devices).
- Sample labeling and ID rubric, redundancy of labeling provisions.
- Sample handling and holding conditions and packaging and shipping from field to lab.
- Chain of custody from point of sampling to final analysis.
- Sampling personnel and/or entity.
- Laboratory identification and location.
- The minimum number of aliquots per event is 6, with a preference for at least 10, for influent and effluent. However, whenever possible it is recommended to obtain a statistically significant number of events as explained at the end of Section 3.2.1.
- All replicate samples (sample duplicates, matrix spike duplicates, LCS duplicates, or other replicates) must be evaluated for a precision criterion not to exceed 20% RPD.

3.1.2.5 Conditions for Abandonment

It is possible that the field site selected will prove to be a poor location for assessing the performance of a typical installation. Basis for abandonment should be identified and agreed upon in advance and can include the following:

- Infestation of animals that damage sampling equipment.
- Unknown hydraulic conditions (such as backwater issues, or unexpected run-on).
- Insignificant pollutant loads entering the MTD rendering inconclusive data.
- A substantial presence of pollutants that are undetected upon entry and/or inlet to the MTD, but are captured and later degrade only to be detected at the outlet (e.g., leaves).
- Unsafe site conditions, including the presence of sanitary sewerage, hazardous traffic, and habitat creation for dangerous animals.

3.2 SAMPLING AND ANALYSIS

The gravimetric index, TSS, has been the most common analytical method to assess runoff in terms of concentration of PM (Rossi et al. 2005). The classification of this suspended PM was originally developed for characterizing wastewater treatment plants' (WWTP) primary clarifiers to quantify those organic solids not significantly separated by a quiescent Imhoff settling process (Metcalf and Eddy 1916), in contrast to those solids separated in 60 min and designated as settleable. These WWTP suspended organic solids have a specific gravity in the range of 1.03 to 1.05 (Tchobanoglous et al. 2003). Thus, wastewater samples collected after the primary sedimentation process of a WWTP do not contain much sand-size and inorganic PM.

A growing number of studies have demonstrated that PM from surface runoff has different properties than those in wastewater with respect to PSD, PM shape, specific gravity, and distribution of pollutants across particle gradations (Makepeace et al. 1995; Sansalone et al. 1998; Sansalone and Cristina 2004b; Li et al. 2005). The granulometric, hydrodynamic, arrival, and temporal characteristics of urban source area rainfall-runoff flows are significantly different from WWTP flows. These rainfall-runoff PM characteristics are not effectively captured by automatic samplers (Furumai et al. 2002; Li et al. 2005) or quantified by standard analytical techniques such as TSS (Gray et al. 2000; Bent et al. 2001; James 2003; Down and Lehr 2005, Selbig et al. 2007). However, automatic samplers and TSS continue to be widely adopted despite reflecting a wastewater-type sampling paradigm and continue to be used as sampling and analytical methods to monitor the chemistry of urban rainfall runoff and examine the transport and treatment of PM despite completely different hydrodynamic and granulometric conditions (Wu et al. 1998; Andral et al. 1999; Drapper et al. 2000; Taebi and Droste 2004; Anta et al. 2006). The sediment concentration methodology recommended in this report is SSC. TSS may also be used as a regional preference.

3.2.1 Storm Tracking and Sampling Intervals

Most verification programs require a percent coverage for an individual storm event. Typically, an 80% coverage is needed. To accomplish this, the field testing organization needs to balance the predicted storm depth and duration and the sampler capacity to achieve the coverage goals.

Whether sampling is conducted manually or automatically, adaptive decision making can be facilitated by knowledge of the seasonal weather patterns and information compiled from local and national radar imagery systems. This information is readily available from different websites, typically based on data from the National Oceanic and Atmospheric Administration (NOAA) or National Weather Service (NWS), and, in many localities, now integrate a high-resolution Google Earth layer and a layer of storm cell vectors. These attributes are public domain even for web-based tools such as www.wunderground.com. Regardless of the method used, it is recommended to set up a recording precipitation rain gauge on the site or as close as possible to it to obtain accurate data on the precipitation of each event. When using data from NOAA or NWS, the monitoring stations might be several miles away from the site and, thus, not give an accurate depiction of the actual stormwater event that triggers the sampling device.

Using real-time radar imagery that is available throughout North America at 3 to 5 min intervals, adaptive sampling decision making can be implemented even within small watersheds (less than 1,000 m^2) where the typical rainfall to watershed runoff residence time is between 5 and 15 min. When a first-order exponential first flush for SSC or TSS is suspected [i.e., load transport during the rising limb of the hydrograph (see Fig. 3.1)], more frequent sampling intervals (1 to 5 min) can be employed. For a simple single hydrograph event, adaptive decision making can be

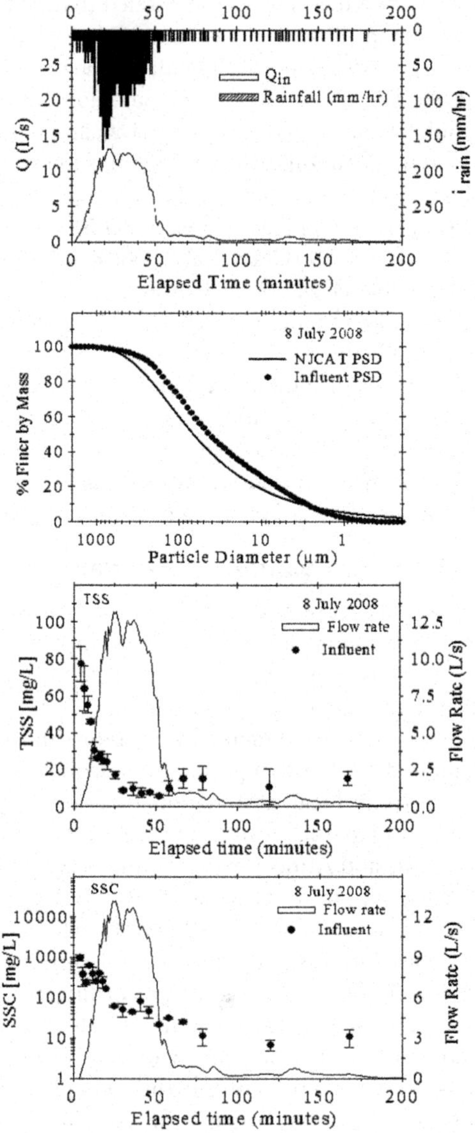

Fig. 3.1. Coupling of hydrologic loads with PM measurements: Mean PSD for the event and transport of TSS and SSC are shown

based on radar imagery to generate coarser sampling intervals during the falling limb of the event, for example 5 to 10 min for the treatment duration of the event. Such adaptive sampling is driven by the lower gradient of concentration on the falling limb. A similar rising/falling limb process can be roughly approximated by programming with volume-weighting sampling intervals automatically. However, the greatest benefit of adaptive decision making with radar imagery and manual sampling is observed in complex, multiple hydrograph events. This process has been used successfully in sampling 40 events (located between Baton Rouge, Louisiana and Gainesville, Florida) without a single failure to fully capture and representatively sample a storm. Current radar imagery can be calibrated for a given locality to minimize mobilization effort when deciding whether to capture an event or not.

To obtain statistically significant data, the *Urban Stormwater BMP Performance Manual* recommends that a minimum number of samples per event be taken based on the following calculation (Roesner et al. 2007):

$$n = \frac{4(\sigma)^2}{E^2} \tag{3-1}$$

where n = number of samples to be taken; σ = standard deviation of the data; and E = allowable error.

It is recommended that a range of possible storm events be monitored to obtain representative data for the site. In some cases these data can be normalized to be more representative of the long-term rainfall distribution (Lenhart and de Ridder 2008). From the study of previous rainfall data, a statistical analysis can be performed to determine what kind of rainfall events occur. The best approach is to sample a sufficient number of average-intensity rainfall events and some light-intensity and high-intensity events. To achieve this goal, it might be necessary to analyze the rainfall patterns in the different seasons. Some studies suggest that a goal of sampling 10 storms per year is appropriate for most humid areas in the US (Law et al. 2008). These Guidelines suggest a minimum of 15 total sampling events so that the rainfall events sampled cover a diverse set of conditions and situations to which the MTD will be subject during operation.

3.2.2 PM Sampling and Analysis

Despite the fact that most MTDs are evaluated on the basis of TSS removal, it is clear that a large fraction of the solids retained in a MTD are not reflected in the mass load calculations associated with TSS removal. Even though most verification programs will continue to focus on TSS (as that is what is being regulated) a field study should also seek to characterize the entire spectrum of PM.

PM analysis categorizes any noncolloidal PM larger than a nominal size of 1 μm except for floatable matter. PM is classified into three fractions: suspended, settleable, and sediment. The term *particulate matter* is preferred as opposed to particle or solids (adopted from wastewater) because PM transported in runoff might not always be in the form of individual discrete particles but can also be aggregated PM as experimentally demonstrated elsewhere (Slattery and Burt 1997). Additionally, the use of PM is consistent with the terminology in the atmospheric literature (Ying et al. 2007).

The entire size gradation of PM is important to characterize, at the very least prior to beginning of testing, maintenance activities, and the end of the test period. There is broad agreement that MTDs fill with coarse PM, gross solids, and floatables that are unaccounted for when only sampling for TSS. There is also general agreement that most MTDs are not effective at separating suspended PM as is evidenced by arguments that there are irreducible levels of effluent that cannot be achieved by MTDs (Strecker et al. 2001; McNett et al. 2011). To characterize the entire gradation of PM and the three basic classes of PM fractions (suspended, settleable, sediment) at each sampling time, a replicate set of samples must be taken either automatically or manually.

3.2.2.1 Sediment Fraction of PM (>75 μm)

The sediment fraction includes all sizes larger than 75 μm (#200 sieve). This is consistent with published classifications that separate *coarse* from *fine* soil particles and silt-size from sand-size particles at 75 μm (ASTM 2002; Rushton et al. 2007). This fraction can include not only coarse sand (inorganic particulates) but also litter (i.e., paper, plastic, metal, glass) and biogenic debris (i.e., leaves, twigs, and grass) up to 4,750 μm.

3.2.2.2 Settleable Fraction of PM (~25 to 75 μm)

While not expected, if possible after prescreening through a #200 sieve, the sieved filtrate is placed in an Imhoff cone. The settleable fraction of PM is defined as the PM that settles in an Imhoff cone in 1 h.

3.2.2.3 Suspended Fraction of PM (1 to 25 μm)

The PM remaining suspended in the Imhoff cone after 1 h of quiescent settling is defined as the suspended fraction. The separation between the settleable and suspended fractions is based on particle properties. The demarcation size between these two fractions can vary depending on specific gravity and shape of settling particles.

It is recommended that the following guidelines be followed to ensure adequate sampling of PM:

1. Sampling for PM should be carried out immediately before and after the MTD (i.e., at both inlet and outlet of the MTD) whether by automated or manual sampling.
2. The sampling configuration should also provide for overflow sampling and flow measurement.
3. Aliquots must be retrieved from the rising limb, peak, and falling limb of the hydrograph. While up to 24 flow-weighted aliquots would be ideal, the

minimum number is 6, provided they are well distributed throughout the event.

4. Sampling intervals can be based on knowledge of the seasonal weather patterns and information compiled from radar imagery systems (Sansalone and Cristina 2004a). It must be noted that on an event scale this is still inadequate.
5. Manual sampling of the full cross section of flow may be conducted for select MTD installations. If an event mean concentration (EMC) alone is not acceptable for a particular protocol, individual sample and sample replicate analysis can be conducted with mathematical compositing based on measured flow volume and time. This is a much more difficult and laborious process, but provides representative results and material balance checks (Sansalone and Kim 2008; Kim and Sansalone 2008). Manual sampling can also be useful when large quantities of oil and grease are expected, because automatic samplers might not fully capture their incidence due to their floatability. If sampling is conducted manually, safety concerns such as lightning, wind, and traffic must be accounted for.

3.3 METHODS

The following is a summary of the water column field monitoring data requirements for PM separation by MTDs.

3.3.1 Initial Conditions Data Component

The parameters that define the hydrology and physical-chemical properties at the MTD system boundaries can be classified as

- **Hydrologic** (rainfall, previous dry hours, wind, air temperature, relative humidity, infiltration, evaporation, watershed storage, and watershed characteristics),
- **Hydraulic** (flow rate, velocity, cross-sectional flow area, wetted perimeter, roughness, shear stress, hydrodynamic regime, temperature, degree of saturation, and residence time distributions), and
- **Particulate** (PSDs, PM measured as SSC, suspended PM, settleable PM, sediment PM, gross solids, floatables, constituent distribution on PSDs, specific gravity, solubility, leaching, MTD sludge properties including depth, density, and scour).

Three levels of boundary/initial conditions field verification requirements are proposed. These Guidelines focus primarily on Level 1:

- **Level 1** field verification requirements are for event-based MTD examinations and commensurate with the field monitoring guidance provided herein for

PM. Level 1 testing is typically performed to meet criteria established by a regulatory program for the verification and ultimate certification of an MTD by a regulatory body. The exact requirements and parameters of the study are typically outlined in the QAPP and guidelines established through programmatic process.

- **Level 2** field verification requirements are more detailed than what are outlined here, requiring PM surface properties, discrete sampling, and leaching characteristics of MTD sludge, and are appropriate for monitoring and modeling event-based MTD examinations and continuous simulations requiring temporal input and output data, and event-based system data.
- **Level 3** requirements are research or specialized datasets for MTD examinations looking at specific pollutants beyond PM and including partitioning: chemical, nutrients, metals, organics, endocrine-disrupting compounds, and microbiological. These datasets are combined with event-based and continuous simulation modeling and water chemistry examinations such as charge, volume, and mass balances.

3.3.2 Level 1 Testing

3.3.2.1 Events Monitored

A minimum of 15 events are required for an MTD certification. Events should be within a two-year period, and event rainfall depth should be sufficient to generate runoff to collect a minimum of six aliquots (10 aliquots is preferred). There should be at least six previous dry hours. The median storm event sampled should be within 20% of the depth of the median precipitation event at that location. Adjustment of this tripping criteria can be made based on site-specific conditions such as temporal variations in rainfall distributions or inflow conveyance geometry restricting accuracy of flow meters. Event mean flows should be reasonably distributed between 25 and 75% of the MTD design flow rate, but variability is preferred.

3.3.2.2 Hydrologic Parameters

- A *watershed delineation* so that a rainfall-runoff model can translate rainfall loadings to runoff loadings at the MTD system location for the watershed of interest. The benchmark model of choice is USEPA Storm Water Management Model (SWMM), which is nonproprietary.
- A *watershed-based and event-based temporal rainfall record* for storm events captured. Representative rainfall depths should be recorded in increments of 0.01 in.
- A measurement of *antecedent dry period (ADP) is needed. Where appropriate (e.g., a biologically based MTD), wind speed and direction, air temperature, evaporation indices, and relative humidity* for the watershed or locality on an

event basis should be collected. ADP is determined at the site of study, while the balance of these data is available from a nearby weather station.

3.3.2.3 Hydraulic Parameters

- *Flow depths and velocities* at increments no greater than 5 min intervals for the entire inflow and outflow duration for each inflow and each outflow of the MTD system.
- *Temperature* is recommended to be measured at increments to match hydraulic flow measurements for the entire inflow and outflow duration of each inflow and outflow of the MTD system, as well as initial storage and final storage temperatures.
- *Pressure (depth) measurements* should be made as a function of volume treated for MTD systems taken online just upstream and downstream of the MTD so head loss across the MTD is measured as a function of flow rate. This is a critical need for porous media systems.

3.3.2.4 Particulate Parameters

The analytical approach used needs to be established based on the regulatory requirements, the data quality objectives, and the practical limit of the sample volume. Based on these needs there should be a hierarchy of analytical needs in the event that the volume needed to complete the suite of analytes exceeds the event sample volume.

Presented below is an idealized approach that accounts for all of the PM associated with the MTD.

- *A measured volume for each PM fraction* should be determined to the nearest 10 mL. Duplicated 4-L influent and 12-L effluent samples should be prepared with the exact volumes recorded.
- *A prescreening through a #200 (75 μm) stainless steel or nylon sieve* of each duplicated 4-L composite sample and 20-L composite sample should be first carried out by passing the entire volume before fractionation at the 0.45-μm level. The sieve should be free of any holes. The grid on the seive should be weighed and gently disaggregated. A mass balance shall be made by summing the mass components on each sieve, allowing no greater than a 5% mass balance error from the original dry weight.
- *An Imhoff cone analysis for settleable (<75 μm) and suspended (<25 μm) PM* of each duplicated 1-L sample. A turbidity measurement taken for each sample before initiating the Imhoff cone analysis. Volumes for this analysis are recorded to the nearest 10 mL.
- *An SSC analysis* for each 500-mL duplicate sample that is paired with a 500-mL duplicate PSD sample. Volumes of each are recorded to the nearest mL, and the PSD sample should contain 600 mL of sample.

- *A PSD analysis for each duplicated and paired PSD sample.* These PSD samples must be matched with their SSC partners to convert % volume of PM by particle size to an absolute gravimetric value in mg/L.
- *An estimation of sludge depth* should be made at the beginning and end of the monitoring. A 2,000 dry grams sludge sample should be taken from the MTD (location determined from modeling or measurement of short-circuiting) for determination of PSD, specific gravity, and contamination levels as a function of PSD.
- *A determination of specific gravity for the sub 75 μm and larger than 75 μm fractions* and percent organic matter for sludge recovered from the MTD.

3.3.2.5 Porous Medium, Soil, or Filter Matrix

- *A representative set of porosities* of a porous medium, whether surrounding soil of an infiltrating MTD or as a filter matrix, should be measured.
- *A representative set of filter or soil media size distributions* should also be measured.
- *Head loss as a function of flow rate for each event* such that filter ripening can be compared with head loss generated by flow rate should be measured.

3.3.2.6 Vegetation

- *An identification of the vegetation species* and approximate size and age should be provided.

3.4 MAJOR COMPONENTS OF FIELD TESTING

Field testing can be used for certification, numerical (or analytical) model development, parametric evaluations, or controlled performance evaluations. Even if similarities among rainfall, runoff, and PM loads are different for testing regions, tools such as frequency distributions and fundamental data collection (as opposed to indices such as an EMC) such as PSDs and the coupling of hydrology with PSDs, allow the transferability of field verification within regional climatic zones. Such verification should be sufficient in scope, depth, and quality that a public database of testing should be available to help discern the accuracy of MTD performance claims.

The 10 major components of MTD field verification are

1. Characterizing the MTD configuration, contributing watershed properties, uses, and conditions.
2. Matching the MTD unit operation hydraulic and volumetric capacity to the watershed loads.

3. Characterizing influent and effluent PSDs, particle density, concentration, and PM fractions.
4. Historical rainfall frequency distributions and event-based rainfall (minimum of 0.10 in.).
5. Hydrodynamics: influent and effluent, flow and velocity, MTD residence time distribution, and head loss.
6. Sufficient event number for statistical/probabilistic evaluation of MTD (i.e., 15–18).
7. Representative sampling and analysis for events across 50% of annual rain depth and mass balances.
8. Quantitative impact of operation and maintenance and scour.
9. Discretization of data to allow for MTD modeling and validation of continuous simulations.
10. Development of computational fluid dynamics coupled with continuous simulation models on an intra-event and longer-term (seasonally or annually) basis to evaluate and validate the sequestration behavior and misbehavior (such as constituent export) from an MTD and the maintenance frequency thereof.

3.4.1 Post-Test and Maintenance Sampling of Herbaceous/ Sediment Matrix

An uncontrollable watershed variable is the mass of vegetative material entering a MTD. This is most easily determined at the end of the sampling period and whenever maintenance takes place. In some watersheds this mass can be orders of magnitude higher than the sediment mass in the water column. Vegetative material can accumulate in the water-filled vaults of BMPs or in dry conditions in screens. Vegetation can be ground up to fine material indistinguishable from sediment with the naked eye. It will also decompose in water becoming SSC/TSS during resuspension in intra-event flows. Leaves trapped in screens will form a fine mesh filter that can trap significant masses of sediments. Therefore, the accumulated herbaceous/sediment material in an MTD should be analyzed to determine the organic/inorganic content of the mass. The inorganic component of the matrix is sediment to be added to the water column EMC masses to give a total sediment mass loading from the stormwater runoff.

Mass build-up is unpredictable, often taking several rain events to accumulate enough material in the bottom of a vault box to take a sample. With screens above the water line, the build-up of leaves during small storms that do not trip autosampler sampling may require more frequent cleaning than autosampling events. Both of these types of conditions can prohibit the matching of water column samples and masses to gross solids sampling and cleaning events. Thus, the calculation of annual mass removal efficiency becomes appropriate for accurately measuring the MTD performance.

3.4.1.1 Sampling Procedures in Wet Chambers

Trash, herbaceous material, and sediment samples from wet MTD chambers will be collected in accordance with SOPs using a stainless steel hand auger or petite ponar. Samples will be volume-weighted and composited into one sample and processed in the lab for chemical analysis.

Total accumulated volume will be calculated by multiplying the horizontal cross-sectional area of each chamber by the average depth of the accumulated material in that chamber. The average depth of accumulation will be determined by measuring the depth of material at a sufficient number of locations to obtain a reasonable average depth estimate for the chamber.

The box chambers will be vacuumed clean after each sediment sampling event by the responsible municipal maintenance entity under the supervision of the researcher and total volume and mass of removed sediment calculated.

Two times during the life of the project, samples of bottom chamber material (trash and debris, herbaceous material, and sediment) will be collected from each baffle box chamber. One sampling event will be performed after 4–5 storm events, depending on material accumulation, and the second at the end of water quality monitoring. At least one sampling event will be performed during or immediately after the *wet season*.

Sample volumes from individual chambers must be large enough to form a composite sample volume that is sufficient for all subsequent analyses. The material in each chamber will be combined to form a single volume-weighted composite sample. Compositing will be done on a weighted basis; the weighing criterion will be based on the volume of herbaceous material and sediment accumulated in each chamber (trash and debris not included). With the weighted compositing procedure, chambers with larger accumulated volumes (typically the upstream chambers) will contribute more material to the composite. The volume-weighted composite samples (with trash and debris removed as much as possible) will be analyzed for sediment content and classified based on Section 3.2.2, "PM Sampling and Analysis." It will not be possible to separate all organic material from the sediment (Rushton et al. 2007). A percent organic matter test (ASTM D29742014) will be used to determine the inorganic sediment component of each stratum of the PSD.

3.4.1.2 Sampling Procedures in Dry Screens

Each time the screens are cleaned, trash, herbaceous material, and sediment trapped in the screens will be removed and sorted. The total volume and mass of removed sediment and the total volume and mass of debris will be calculated. Total accumulated volume will be estimated by averaging horizontal cross section and depth measurements. Prior to cleaning, the screens should be photographed and estimates of the remaining percentage of open area should be recorded.

Two times during the life of the project, the material trapped on the screens (trash and debris, herbaceous matter, and sediment) will be completely removed and subsampled for laboratory analysis. One sampling event will be performed after 4–5 storm events depending on material accumulation and the second at the end of water quality monitoring. At least one sampling event will be performed during or immediately after the wet season.

The material removed from the screens will be spread out on a flat surface. Trash and debris (litter, paper, plastic, metal, glass, cups, styrofoam, bags, branches, 2 × 4, etc.) will be visually identified and removed by hand. The bulk volume of the removed material will be measured and the dry weight determined.

The remaining material (herbaceous material and sediment) will be mixed and dried for 48 h outside, spread out to a thickness less than ½ in. and dried for an additional 72 h before mixing again. At least five samples collected from different locations will be combined into a single composite sample. The individual sample volumes must be large enough to form a composite sample volume that is sufficient for all subsequent analyses. The volume-weighted composite samples (with trash and debris removed as much as possible) will be analyzed for sediment content according to the classification outlined in the Section 3.3.2, "PM Sampling and Analysis." It will not be possible to separate all organic material from the sediment (Rushton et al. 2009). A percent organic matter test (ASTM D2974) will be used to determine the inorganic sediment component of each stratum of the PSD.

References

Andral, M. C., Roger, S., Montrejaud-Vignoles, M., and Herremans, L. (1999). "Particle size distribution and hydrodynamic characteristics of solid matter carried by runoff from motorways." *Water Environ. Res.*, **71**(4), 398–407.

Anta, J., Pena, E., Suarez, J., and Cagiao, J. (2006). "A BMP selection process based on the granulometry of runoff solids in a separate urban catchment." *Water SA*, **32**(3), 419–428.

ASTM. (2002). "Standard practice for classification of soils for engineering purposes." *D2487-06*, West Conshohocken, PA.

ASTM. (2014). "Standard test methods for moisture, ash, and organic matter of peat and other organic soils." *ASTM D2974-14*, West Conshohocken, PA.

Bent, G. C., Gray, J. R., Smith, K. P., and Glysson, G. D. (2001). "A synopsis of technical issues for monitoring sediment in highway and urban runoff." <http://www.co.monmouth.nj.us/health/environmental/303DTMDL/documents/tssssc.pdf> (Sep. 13, 2007).

Down, R. D., and Lehr, J. H. (2005). *Environmental instrumentation and analysis handbook*, Wiley, Hoboken, NJ.

Drapper, D., Tomlinson, R., and Williams, P. (2000). "Pollutant concentrations in road runoff: Southeast Queensland case study." *J. Environ. Eng.*, **126**(4), 313–320.

Furumai, H., Balmer, H., and Boller, M. (2002). "Dynamic behavior of suspended pollutants and particle size distribution in highway runoff." *Water Sci. Technol.*, **46**(11–12), 413–418.

Gray, J. R., Glysson, G. D., Turcios, L. M., and Schwarz, G. E. (2000). *Comparability of suspended-sediment concentration and total suspended solids data*, U.S. Dept. of the Interior, USGS Information Services, Reston, VA.

James, R. B. (2003). "TSS—A viable measure of storm water pollutants?" <http://www.stormwaterauthority.org/assets/55tss.pdf> (Jun. 23, 2007).

Kim, J.-Y., and Sansalone, J. J. (2008). "Event-based size distributions of particulate matter transported during urban rainfall-runoff events." *Water Res.*, **42**(10–11), 2756–2768.

Law, N. L., Fraley-McNeal, L., Cappiella, K., and Pitt, R. (2008). *Monitoring to demonstrate environmental results: Guidance to develop local stormwater monitoring studies using six example study designs*, Center for Watershed Protection, Ellicott City, MD.

Lenhart, J., and deRidder, S. (2008). "Weighting of field monitoring data with probability distributions of daily rainfall." *ASCE/EWRI World Environmental and Water Resources Congress*, ASCE, Honolulu, HI.

Li, Y. X., Lau, S. L., Kayhanian, M., and Stenstrom, M. K. (2005). "Particle size distribution in highway runoff." *J. Environ. Eng.*, **131**(9), 1267–1276.

Makepeace, D. K., Smith, D. W., and Stanley, S. J. (1995). "Urban stormwater quality—Summary of contaminant data." *Crit. Rev. Environ. Sci. Technol.*, **25**(2), 93–139.

McNett, J. K., Hunt, W. F., and Davis, A. P. (2011). "Influent pollutant concentrations as predictors of effluent pollutant concentrations for Mid-Atlantic bioretention." *J. Environ. Eng.*, **137**(9), 790–799.

Metcalf, L., and Eddy, H. P. (1916). *American sewerage practice*, 2nd Ed., McGraw-Hill, New York.

Roesner, L. A., Pruden, A., and Kidner, E. M. (2007). *Improved protocol for classification and analysis of stormwater-borne solids*, Water Environment Research Foundation, Alexandria, VA.

Rossi, L., Krejci, V., Rauch, W., Kreikenbaum, S., Fankhauser, R., and Gujer, W. (2005). "Stochastic modeling of total suspended solids (TSS) in urban areas during rain events." *Water Res.*, **39**(17), 4188–4196.

Rushton, B., England, G., and Smith, D. (2007). "ASCE guidelines for monitoring stormwater gross pollutants." <http://www.stormwater.ucf.edu/conferences/9thstormwaterCD/documents/ASCEguidelines.pdf> (Sep. 13, 2007).

Sansalone, J. J., and Cristina, C. M. (2004a). "First flush concepts for suspended and dissolved solids in small impervious watersheds." *J. Environ. Eng.*, **130**(11), 1301–1314.

Sansalone, J. J., and Cristina, C. M. (2004b). "Prediction of gradation-based heavy metal mass using granulometric indices of snowmelt particles." *J. Environ. Eng.*, **130**(12), 1488–1497.

Sansalone, J. J., Koran, J. M., Smithson, J. A., and Buchberger, S. G. (1998). "Physical characteristics of urban roadway solids transported during rain events." *J. Environ. Eng.*, **124**(5), 427–440.

Sansalone, J., and Kim, J. Y. (2008). "Transport of particulate matter fractions in urban source area pavement surface runoff." *J. Environ. Quality*, **37**(5), 1883–1893.

Selbig, W. R., Bannerman, R., and Bowman, G. (2007). "Improving the accuracy of sediment-associated constituent concentrations in whole storm water samples by wet-sieving." *J. Environ. Qual.*, **36**(1), 226–232.

Slattery, M. C., and Burt, T. P. (1997). "Particle size characteristics of suspended sediment in hillslope runoff and stream flow." *Earth Surf. Proc. Land.*, **22**(8), 705–719.

Strecker, E. W., Quigley, M. M., Urbonas, B. R., Jones, E. J., and Clary, J. K. (2001). "Determining urban storm-water BMP effectiveness." *J. Water Resour. Plann. Manage.*, **127**(3), 144–149.

Taebi, A., and Droste, R. L. (2004). "First flush pollution load of urban stormwater runoff." *J. Environ. Eng. Sci.*, **3**(4), 301–309.

Tchobanoglous, G., Burton, F. L., Stensel, H. D., and Metcalf & Eddy. (2003). *Wastewater engineering: Treatment and reuse*, 4th Ed., McGraw-Hill, Boston.

Wu, J. S., Allan, C. J., Saunders, W. L., and Evett, J. B. (1998). "Characterization and pollutant loading estimation for highway runoff." *J. Environ. Eng.*, **124**(7), 584–592.

Ying, G. X., Ma, J., and Xing, Y. (2007). "Comparison of air quality management strategies of PM10, SO2, & NOx by an industrial source complex model in Beijing." *Environ. Prog.*, **26**(1), 33–42.

CHAPTER 4
Field Data Analysis and Performance Evaluation

Data Evaluation Subcommittee

The focus of this chapter is to address the data analysis and performance evaluation of field data. The existing statistical data analysis and performance evaluation methods that could potentially be used for manufactured treatment devices (MTDs) have been identified. Special emphasis has been placed on data distribution and the issue of normality because it will influence the selection of a suitable data analysis approach. It has been concluded that stormwater data are generally not normally distributed and are frequently lognormally distributed instead. Methods to evaluate performance of MTDs are also examined. In general, the performance of MTDs under laboratory conditions can be determined through the conventional influent/effluent concentration or mass balance basis. However, the performance of MTDs under field conditions can be determined by other methods that may be more practical for regulatory compliance. Among the methods, the effluent probability plot and performance expectation function have been found to be versatile and useful tools to assess a wide range of stormwater best management practices (BMPs), including MTDs.

4.1 POTENTIAL DATA ANALYSIS APPROACHES

The central goal of an MTD verification program is to collect scientifically defensible data in fulfillment of the data quality objectives, quality assurance project plan requirements, and verification objectives. Statistical methods may be used in the design of the laboratory, pilot, or field testing to determine data quality objectives (DQOs). DQOs are an important and useful step to ensure that the data produced by laboratory or field testing are sufficient, both in terms of quality and quantity, to establish meaningful confidence levels for the MTD performance evaluation.

4.1.1 Data Quality and Reporting Statistical Analysis

The use of standardized procedures to collect and report MTD performance data is essential for a successful verification program—not just from one MTD to another, but also from project to project, region to region, and year to year. If consistent methods are not applied, it may be difficult to perform data comparisons, and the value of the data may be diminished. For instance, flow can vary significantly throughout a runoff event, with varying rainfall intensity and runoff volume (Kayhanian et al. 2003; Kayhanian et al. 2007). Flow-proportioned composite samples are therefore considered to be the most representative sampling regimen for field monitoring. The preferred method for water quality evaluation in BMPs is to employ automated monitoring equipment to collect an equal sample volume (aliquot) for every increment of a preset runoff flow volume. The composite sample ideally needs to cover the full event hydrograph with a minimum number of aliquots to account for variation in flow and/or runoff quality throughout the runoff event. However, in some sampling protocols the compositing of grab samples is needed for analyzing large volumes of solid samples (including sediment and litter) where automatic samplers may be impractical (Rushton et al. 2007).

Regardless of the sampling method, the verification protocol should provide a series of steps to be taken both in the field and in the lab to ensure the overall quality of data. The data reporting protocols discussed in the previous chapter help ensure that data from all tested BMPs are reported in a consistent format and that the data records will include sufficient information to permit their full use within a common database. Standardized names and units for the data reported will facilitate data extraction and analysis. In addition, a well-designed database with a user-friendly interface could provide ease of data storage and retrieval for a successful verification program. A schematic of the prerequisites for a successful MTD performance evaluation is shown in Fig. 4.1.

Performing detailed statistical analysis and performance evaluation of MTDs will require additional data collection such as flow, physical characteristics of the manufactured device, site characteristics, and storm event characteristics. Fields and content guidelines for each of these categories have previously been described in the Field Testing chapter.

4.1.2 Statistical Analysis

An important issue to be addressed prior to performing statistical analysis is the distribution of the data. Frequently, data sets from field studies are small and not normally distributed. The data should be either transformed to a normal distribution or subjected to nonparametric methods to address specific statistical questions. One method to verify normal distribution of the data is to create a probability plot of the original and of the log-transform of stormwater runoff data. Normality can also be tested applying the Shapiro-Wilk W test to verify if the data fit a normal distribution. Table 4.1 shows an example of the Shapiro-Wilk W test.

Table 4.2 summarizes the recommended statistical approach to address some of the questions that arise when performing statistical analysis of the data collected from testing stormwater BMPs. Further recommendations of statistical analyses for stormwater control measure (SCM) performance monitoring can be

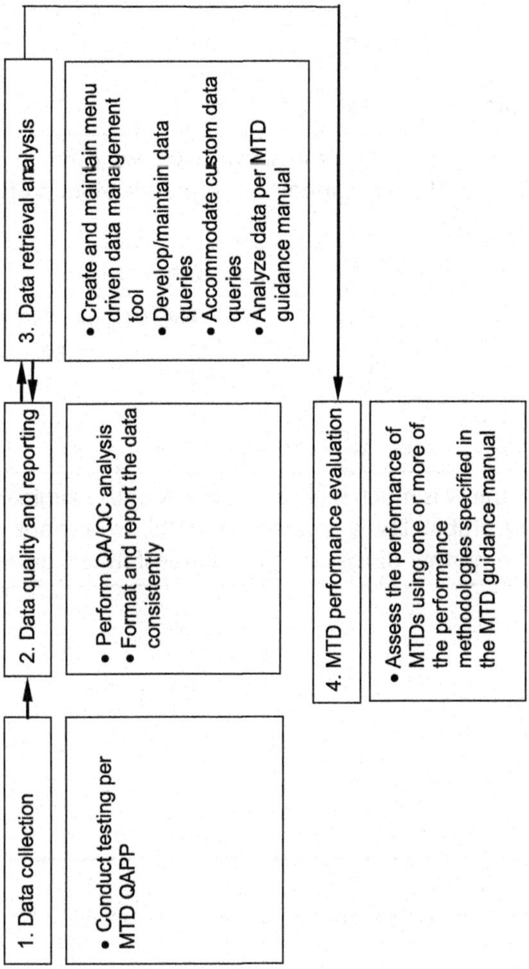

Fig. 4.1. Prerequisites of a successful MTD performance evaluation

Table 4.1. Example of the Shapiro-Wilk W test

Test #	*BMP Effluent Concentration*	*Shapiro-Wilk W Test Result*			
1	10.9	Parameter estimate prior transformation:			
2	4.3	*Parameter*	*Estimate*	*Lower 95%*	*Upper 95%*
3	6.9	μ	11.84	8.75	14.94
4	0.7	σ	13.90	12.03	16.47
5	2.9	W test result:			
6	79.6	W = 0.642	Prob < W = <0.0001		
7	9.6	Because the probability is less than 0.05, we can			
8	4.0	conclude that the assumption of normal distribution is			
9	4.8	not reasonable			
10	14.3	Parameter estimate after log transformation:			
11	14.5	*Parameter*	*Estimate*	*Lower 95%*	*Upper 95%*
12	4.4	μ	1.99	1.77	2.22
13	12.2	σ	1.02	0.88	1.21
14	13.5	W test result:			
15	8.6	W = 0.974	Prob < W = <0.105		
16	8.4	Because the W is much higher that the W of the original			
17	11.7	data and probability is greater than 0.05, we cannot			
18	9.7	reject the normal distribution of logtransformed data			
19	17.8				
20	3.8				
21	4.6				
22	11.6				
23	3.6				
24	1.5				
25	10.3				
26	16.0				

found in the "Urban Stormwater BMP Performance Monitoring" manual (USEPA et al. 2009).

4.2 MTD PERFORMANCE EVALUATION

4.2.1 Review of the Historical Approaches

Numerous methods have been applied to evaluate the field performance of conventional constructed stormwater BMPs and MTDs (see Table 4.3). Among

Table 4.2. Summary of statistical approaches in MTD testing

Statistical Topic	*Question to Be Addressed*	*Statistical Approach*
Sample size	How many samples are needed to achieve desired confidence? How to decide if more samples than initially planned are needed?	USEPA statistical methods for practitioners: EPA QA/G-9 S
Selection of an appropriate statistical test	How to verify if the data are normally distributed? How to verify that the data variability of two or more groups is similar?	Graphical data presentation using histogram, box plot, normal probability plots. Numerical evaluation through Shapiro-Wilk W test
Addressing nondetect in dataset	How to account for nondetect results?	Regression on order statistics (ROS)
Examining outlier in dataset	How to prepare graphical and numerical data summaries to identify potential outliers?	Histogram, box plot, standard and logtransformed probability plots, and scatter plot
Probability of meeting water quality requirements	How to estimate how often the average SCM effluent concentration would meet a water quality standard or regulatory discharge limit? How to estimate the SCM percentage removal of a pollutant with a specified confidence level?	This should be available directly from the effluent probability plot of laboratory or field testing data, assuming the field data sample is large enough to be representative

(Continued)

Table 4.2. Summary of statistical approaches in MTD testing (Continued)

Statistical Topic	*Question to Be Addressed*	*Statistical Approach*
SCM effectiveness for specific pollutant removal	In an influent-effluent approach or before-after approach, how to decide if a given SCM is effective in removing a pollutant?	Hypothesis testing through null hypothesis (H_0) or alternative hypothesis (H_A). Statistical tests include student's t-test with equal variance and t-test with unequal variance. Both tests assume a particular probability distribution
Comparing effectiveness of two or more BMPs	How to compare the effectiveness of two or more pilot BMPs under similar test conditions or under a given geographic location?	Paired t-test comparing the mean of effluent concentration for normal or lognormally distributed data. Comparing three or more BMPs use the analysis of variance (ANOVA). Specific analyses include parametric ANOVA (equal variances), Welch ANOVA (unequal variance)
Developing a linear regression equation	How does SCM effectiveness vary as a function of factors such as storm event characteristics, SCM design variables, operational maintenance practices?	Multiple linear regression using the following form: $y = \beta_0 + \beta_1 x_1 + \beta_2 x_2 + \ldots \beta_m x_m + \varepsilon$ where y = response variable, x = model variable, β = model parameters, and ε = random error
Effectiveness of BMPs over time	How to tell if the effectiveness of an SCM is changing over time within a year or multiple years?	Graphic representation using time-series plot or Mann-Kendall test

Note: Refer to Kayhanian et el. (2002) and Shumway et al. (2002) for data analysis with data below detection limit using ROS the methodologies presented in Table 4.3, influent and effluent concentration or load basis are the most common. There is no definitive methodology; the usefulness of any chosen method depends on the size and type of the data set.

Apart from the effluent probability plot, the statistical methods listed have been used extensively for all types of stormwater BMPs including MTDs.

4.2.2 Effluent Probability Method

The effluent probability method is a visually comprehensive method that can be used to evaluate the performance of MTDs under field conditions. This method is most useful when adequate data have been collected in the field to enable a performance probability plot. Under this approach, a standard probability plot is prepared for both inflow and outflow pollutant event mean concentrations (EMCs) or loads. Before preparing the probability plot, a nonparametric (or parametric when possible) statistical test should be conducted to determine if the difference between influent and effluent is statistically significant. Log normality should also be checked; the probability plot may be prepared based on log transformed data if log normality is found to fit the data. If the log transformed data deviate significantly from normality, other transformations can be explored to fit the data distribution. A clear difference between influent and effluent event mean concentration plots indicates the MTD is performing well. When the difference between influent and effluent EMC is too small to detect, the removal efficiency will be very low or nonexistent. An example probability plot for MTD influent and effluent total suspended solids (TSS) EMC is shown in Fig. 4.2. As shown, it is possible for influent and effluent probability plots to vary with influent TSS concentrations.

A weakness in the probability method is its lack of relationship with direct influent and effluent measurements. Ideally, the performance of the MTD will remain within a narrow range with high probability of meeting effluent requirements regardless of influent concentration or testing condition. For example, the probability plots shown in Fig. 4.2 could be used to determine if the MTD can meet certain regulatory or standard effluent quality requirements. The probability plot shown in Fig. 4.2 indicates that only under ideal performance conditions can an effluent TSS concentration of 30 mg/L be met at all times. Multiple probability plots can also be produced under a single plot to compare the performances of two or more BMPs.

Another problem with testing and analyzing MTD performance is that pollutant removal efficiency tends to decrease as influent concentration decreases. The baseline concentration below which an MTD may not provide treatment is commonly considered to be 20 mg/L TSS, although removal at lower concentrations has been observed. In many cases this irreducible baseline concentration does not affect the system performance. The point where MTD performance starts to decrease is often close to the point where the influent concentration is low enough that it does not exceed the discharge requirement and hence a high level of treatment is not needed. The baseline concentration poses a regulatory problem when percent removal is the only criterion used. For example, an 80% TSS removal

Table 4.3. Summary of methodologies used for MTD performance evaluation

Method	*SCM Performance Evaluation Method*	*Approach/Formula*	*Applicable Verification*
1	Pollutant EMC reduction for single test or storm event	$\text{Efficiency} = 100 \times \left(1 - \frac{\text{Effluent EMC}}{\text{Influent EMC}}\right)$	ETV, TARP, TAPE, EPA
2	Pollutant load reduction for single test or storm event	$\text{Efficiency} = 100 \times \left(1 - \frac{\text{Effluent load}}{\text{Influent load}}\right)$	ETV, TARP, TAPE, USEPA/ASCE
3	Pollutant load reduction for multiple tests of storm events	$\text{Efficiency} = 100 \times \left(1 - \frac{B}{A}\right)$	ETV, TARP, TAPE, USEPA/ASCE
4	Mean pollutant concentration removal efficiency	$\text{Efficiency} = 100 \times \left(1 - \frac{\text{Average effluent conc.}}{\text{Average influent conc.}}\right)$	USEPA/ASCE BMPP perfor. monitoring
		This technique does not require flow measurement, and it might be useful for pollutants with grab sampling where no flow-weighted data are available	
5	Regression of loads (ROL)	Compute effluent and influent load same as method 2; perform linear regression based on outflow and inflow load and determine the R^2 and slope of regression line. Under special circumstances, the regression lines may be forced through (0,0) as it may produce a more realistic load removal	USEPA/ASCE BMP perfor. monitoring

6	Relative to achievable	$\text{Efficiency} = 100 \times \left(1 - \frac{\text{Attainable limit conc.}}{\text{Average influent conc.}}\right)$	USEPA/ASCE BMPP perfor. monitoring
7	Relative to water quality standard	$\text{Efficiency} = 100 \times \left(\frac{\text{Avg. infl. conc.} - \text{avg.effl. conc.}}{\text{Avg. infl. conc.} - \text{WQ standard conc.}}\right)$	USEPA/ASCE BMPP perfor. monitoring
8	Effluent probability method	This method provides a strong visual SCM performance evaluation by presenting the standard parallel probability plot of both influent and effluent EMCs for all storm events or performance tests	USEPA/ASCE BMPP perfor. monitoring

Effluent load = effluent EMC × effluent volume

Influent load = influent EMC × influent volume

$$B = \left[\sum (\text{effluent EMC}_1 \times \text{flow } V_1) + (\text{effluent EMC}_2 \times \text{flow } V_2) + \ldots\ldots\ldots(\text{effluent EMC}_n \times \text{flow } V_n)\right]$$

$$A = \left[\sum (\text{influent EMC}_1 \times \text{flow } V_1) + (\text{influent EMC}_2 \times \text{flow } V_2) + \ldots\ldots\ldots(\text{influent EMC}_n \times \text{flow } V_n)\right]$$

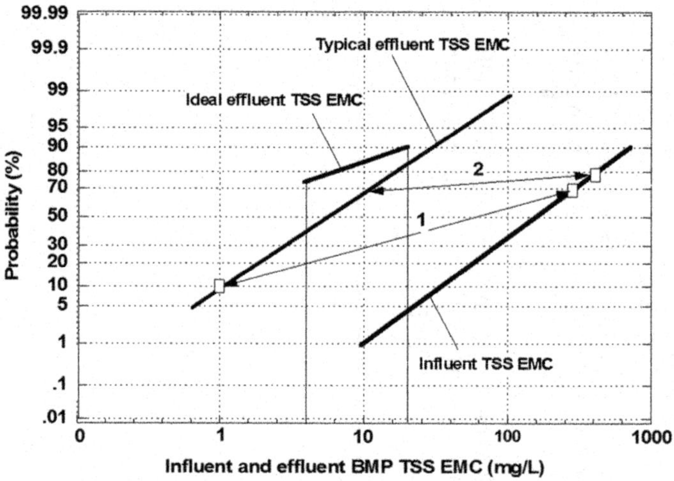

Fig. 4.2. Example probability plot for MTD influent and effluent TSS EMC

regulation would require an effluent concentration of 4 mg/L from a 20 mg/L influent concentration. In cases where most pollutants are associated with smaller particle sizes, some MTDs may not meet the discharge load requirement at full capacity flow rate. There is value in requiring MTDs to report performance even at low concentrations, recognizing that removal may not be as important if discharge limits are being met. At the other extreme, discharge limits may be exceeded at very high influent concentrations even though the MTD is capable of removing 80% of influent concentration. Therefore, testing the MTDs under a wide range of influent flow and of pollutant concentration is essential to determine the true performance characteristics.

4.2.3 SCM Performance Expectation Functions

A common standard used by regulatory agencies for certification of MTDs is an 80% TSS removal rate, or in some cases 80% of total annual TSS load reduction. These requirements do not fully encompass how MTDs perform in the field. A flat effluent standard for BMPs (e.g., 20 mg/L TSS effluent) has been suggested as an alternative. However, the performance of a SCM is probabilistic and presumptive, and it is not practical to set performance levels based exclusively on effluent concentration. Having a method that combines effluent performance standards and evaluation of MTD performance from field data is ideal. A performance expectation function (PEF) can achieve these goals.

The regulatory agency will be responsible for defining the PEF based on its specific water quality goals. The agency will define the baseline effluent concentration. For influent concentrations above the threshold, percent removal will be used. For example, with a baseline concentration of 20 mg/L, an agency would set an effluent guideline of 20 mg/L for an influent concentration of 100 mg/L or less. For concentrations greater than 100 mg/L, the performance expectation is 80% removal efficiency.

Figure 4.3 illustrates the application of PEF in two ways. The first is a plot of influent concentration versus percent removal, and the second is a plot of influent versus effluent concentration. These graphs represent the performance expectation for the MTD. Because the MTD performance is probabilistic it is expected that some of the data points will be above the line and some will be below the line. Once a PEF is defined, field and laboratory testing performance data can be used to compare it with the expected performance curves.

A method of data comparison is the sign test. This is a simple nonparametric statistical test to estimate if the scatter of the points above the line represents the same population or a population that rests above or below the line. For example, if the MTD performance characteristic follows the PEF, it is reasonable to expect that 50% of the points are located above the line and 50% below. If higher frequencies of occurrence at high confidence limits are observed above or below the line, this may suggest that the MTD is outperforming or underperforming the expectation. The sign test may be of use when it is only necessary to know if observed differences between two conditions are significant.

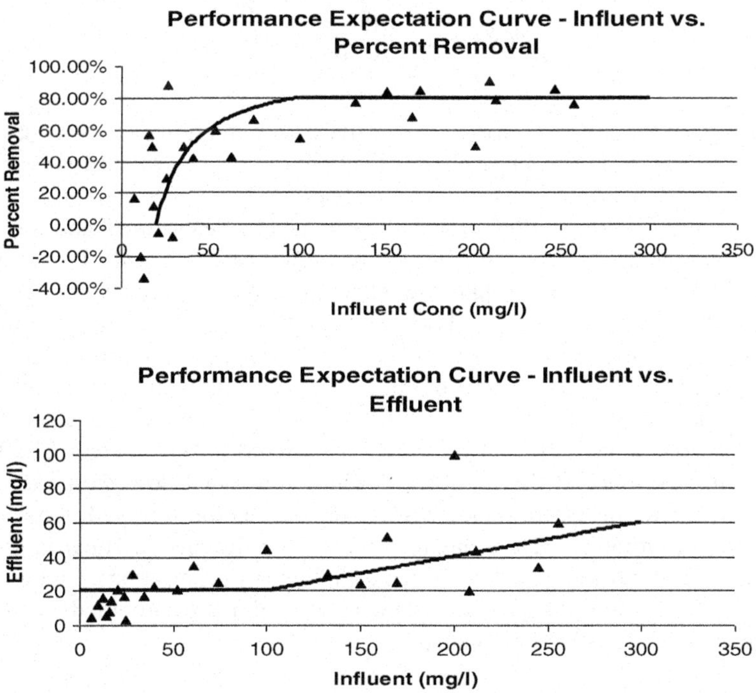

Fig. 4.3. Example TSS performance expectation functions versus observed data

In a sign test, the concentration differences are calculated by subtracting the observed values from the expected ones. Positive numbers are then assigned a plus sign, and negative numbers are assigned a negative sign. Differences of zero (i.e., Observed = Expected) are omitted. The outcome of the number of points above and below the line is compared with a population when it is expected that half the points are above the line and half are below. Using a binomial distribution, the probability that the number of occurrences above (or below) the line is calculated. The probability is then evaluated to decide if the samples represent the PEF. There are three possible outcomes for this test:

1. The probability of the observed data matching the expected values is high.
2. The probability of the observed data not matching the expected values is high. The observed values are larger (+) than the expected values.
3. The probability of the observed data not matching the expected values is high. The observed values are lower (–) than the expected values.

Outcomes 1 and 2 suggest that that the MTD might meet or exceed the expectations on the basis of concentration. Outcome 3 suggests that the MTD does not comply with the defined PEF and should not be approved. The probability of success for a given number of trials meeting the expectation can be formulated as

$$P(X) = \frac{n!}{(n-X)!X!} p^X q^{n-X} \tag{4-1}$$

where $P(X)$ = probability of success (+) in n trials; p = numerical probability of success (use $p = 0.5$); q = numerical probability of failure [use $q = (1 - p) = 0.5$]; n = number of trials, and X = number of successes (positives).

The cumulative binomial probability can also be determined, and the objective cumulative numerical value of 50% or higher can be specified. Because the binomial distribution has a high variability it is possible to define an acceptable range within which the device will still be in compliance (for example if more than 40% of the test results are above the curve). A device can also perform better than the expected PEF for some ranges of influent concentration and worse in others. It is for the regulatory agency to define the acceptable parameters. The results of a Wald-Wolfowitz test (also called a *runs test*) can aid in this process. The Wald-Wolfowitz test looks at the number of values above or below the performance curve or below the performance curve with data points ordered according to influent concentration. If the obtained value (the number of runs) is *low*, the device is not performing uniformly across influent concentrations. To perform a Wald-Wolfowitz test, the mean and standard deviation are computed and then an approximate Z-statistic is calculated

$$(R - \mu_R)/\sigma_R \quad \text{(rejecting for values below } -2\text{)}$$

If there are n observations above the performance curve and m observations below (for a total of $n + m = N$) and if there are R runs, then

$$\mu_R = 1 + \frac{2mn}{N} \tag{4-2}$$

$$\sigma_R = \sqrt{\frac{2mn(2mn - m - n)}{N^2(N-1)}} \tag{4-3}$$

An MTD load reduction evaluation can also be determined using both the expected load removal (expected concentration times the actual runoff volume) and the observed load removal. The difference between these two values represents a residual that can be further analyzed. If the net sum of the observed mass removal is greater than expected, then the mass load basis will be satisfied as well.

The objective of the PEF approach is to allow an agency to define performance criteria that are consistent with how MTDs operate in the field. It allows analysis of both percent removal and load reduction. It has the potential of providing a method of evaluation that can be defined and agreed upon between both the regulatory agency and the manufacturer.

References

Kayhanian, M., Singh, A., and Meyer, S. (2002). "Impact of non-detects in water quality data on estimation of constituent mass loading." *Water Sci. Technol.*, **45**(9), 219–225.

Kayhanian, M., Singh, A., Suverkropp, C., and Borroum, S. (2003). "The impact of annual average daily traffic on highway runoff pollutant concentrations." *J. Environ. Eng.*, **129**(11), 975–990.

Kayhanian, M., Suverkropp, C., Ruby, A., and Tsay, K. (2007). "Characterization and prediction of highway stormwater pollutant event mean concentrations." *Environ. Manage.*, **85**(2), 279–295.

Rushton, B., England, G., and Smith, D. (2007). "Proposed guidelines for monitoring stormwater gross solids." *World Environmental and Water Resources Congress*, ASCE, Reston, VA.

Shumway, R., Azari, R., and Kayhanian, M. (2002). "Statistical approaches to estimating mean water quality concentrations with detection limits." *Environ. Sci. Tech.*, **36**(15), 3345–3353.

USEPA (U.S. Environmental Protection Agency). (2000). "Guidance for data quality assessment practical methods for data analysis." *EPA QA/G-9, QA00 UPDATE, EPA/600/R-96/084*, Washington, DC.

USEPA, ASCE, Geosyntec Consultants, and Urban Drainage and Flood Control District. (2009). "Urban stormwater BMP performance monitoring." *EPA-821-B-02-001*, Washington, DC.

CHAPTER 5

Data Reporting

Data Reporting Subcommittee

Data acquired through laboratory and field testing require detailed reporting. The certification process itself can arguably be as difficult as the device testing process. Because of a lack of reporting guidance for the dominant testing protocol, vendors are left to discover the reporting requirements. This can lead to an inefficient, incomplete, and multistepped review process that can take years to negotiate effectively. The result of effective data reporting is a system performance evaluation report (SPER) that presents sufficient information to determine if a manufactured treatment device (MTD) meets the requisite performance standards within a reasonable level of statistical confidence. A SPER includes both summary data and sufficient experimental data to provide a detailed review of system performance. A critical review of leading testing programs was performed and used as the basis for these Guidelines. The first SPER for an MTD was also included in the review.

5.1 BACKGROUND

Prescriptive testing protocols are needed because the range of influences upon testing and performance data are wide and not always well understood. Influences include watershed characteristics, storm event characteristics, maintenance requirements, system sizing, and testing methodologies. A testing protocol that does not dictate methods introduces biases that are challenging to account for. In addition, for performance characteristics to be compared across a range of different MTDs, similar analytical methods and monitoring strategies must be used. The function of standard protocols is to provide both a uniform testing environment and an accelerated review process that aids in the acceptance of stormwater technologies.

If the testing protocol adequately addresses these issues, the testing results must be presented in a clear and simple way so a committee of scientists is not necessary to synthesize and interpret the results. The staff of the many regulatory agencies that will be reviewing the test results cannot be expected to decipher the range of factors influencing system performance. Reporting results require sufficient information ranging from detailed appendices of experimental data to summary data indicating if the systems met requisite performance standards within a reasonable level of statistical confidence.

A consistent reporting format does not currently exist, which has led to a nontransparent and inefficient application and review process. Consistent and thorough data reporting requirements will aid regulatory review and vendor application alike.

5.2 REPORTING FRAMEWORK

The reporting framework includes information ranging from the testing location, MTD-specific information, test methods and procedures, and testing and sampling conditions. Some reporting items are unique to either field versus laboratory studies. The performance results and discussion are presented from two perspectives: (1) complete testing summary statistics and (2) on an individual event basis. Overall system analysis is needed to determine basic MTD effectiveness. Such analysis includes a range of performance metrics: removal efficiencies, efficiency ratios, and a performance expectation function (Lenhart 2007). The use of removal efficiency was included because of its simplicity and its widespread presence in regulation. A performance expectation function is proposed to address low influent contaminant concentrations with respect to changes in system performance. Performance statistics are used to provide deeper insight into system behavior and include components of the International Stormwater Best Management Practices (BMP) Database (BMPDB) (ASCE et al. 1996). Individual storm reports (ISRs) are used to examine system performance and variability with respect to individual storms and event characteristics. Existing reporting protocols for the Technology Acceptance Reciprocity Partnership, or TARP (2003), and Technology Assessment Protocol–Ecology, also known as TAPE (Ecology 2008), programs were used as guides for the reporting framework in Table 5.1.

An example of a verification report can be viewed online through the New Jersey Department of Environmental Protection (NJDEP) at http://www.state.nj.us/dep/dsr/bscit/NJCAT_Fnl_Ver(Stormfilter).pdf. This NJCAT 2007 report has been approved by the New Jersey Corporation for Advanced Technologies and NJDEP. The proposed framework is mostly consistent with recommendations provided in the previously mentioned report.

5.3 INTERNATIONAL STORMWATER BMP DATABASE

The reporting framework was broadened to include data reporting criteria to meet the requirements of the BMPDB. The purpose of the BMPDB is to provide scientifically sound information for improved system design and selection and

Table 5.1. Data reporting framework for MTD certification testing

Item	*Reporting Elements*
1.	Summary: Executive summary with rated performance rating, study summary, data collection summary
2.	Definitions
3.	Site Conditions: longitude, latitude, land cover type, land use activities, site conditions, site elevations and slopes, location of sampling equipment, location of on-site stormwater collection system, and a description of any upstream BMPs
4.	Technology Description:
	a) The specific device used (model number, size, operating rate or volumetric flow rate)
	b) Functionality of treatment mechanisms including pretreatment and bypass requirements
	c) Physical description: engineering plans, site installation requirements
	d) Sizing methodology used for test: either manufacturer's sizing methodology or agency-specific sizing requirements (flows, volumes, runoff depth, etc.)
	e) Maintenance procedures
5.	Test Methods and Procedures:
	a) Particle size for influent, effluent, and residuals; mass based, concentration based
	b) Water quality parameters monitored
	c) Data quality objectives (DQO), quality assurance (QA) methods, and measurement accuracy for the observations
	d) Measuring instruments, sampling frequency, and sampling program information
	e) Sampling locations and peak concentration timing
6.	Testing and Sampling Event Characteristics:
	a) Storm date, depth, antecedent dry period, intensity, duration, season, type of runoff (precipitation, snowmelt, groundwater, etc.)
	b) Number of influent and effluent aliquots; storm volume, percent storm treated influent, effluent, and peak flow rate
	c) Comparisons with DQOs
	d) System timeline (start and completion, sample events, rainfall events, maintenance occurrence)
	e) Water level within system and rainfall for testing duration

(Continued)

Table 5.1. Data reporting framework for MTD certification testing (Continued)

Item	*Reporting Elements*
7.	MTD Performance Results and Discussion:
	a) Event mean concentrations for influent and effluent with summary statistics (*N*, mean, median, coefficient of variation, standard deviation, one-tailed sign t-test)
	b) Detection limits and confidence intervals
	c) Performance metrics: removal efficiency for EMC and mass loads, efficiency ratio, performance expectation function
	d) Statistical evaluation: time series plot, box, and whisker with confidence intervals; effluent probability method; linear regression
	e) Solids characterization: influent, effluent, residuals particle size analysis
	f) Accumulated mass reductions
	g) Individual storm reports with event characteristics (6a and 6b), combination event hydrograph and hyetograph with sample times, system performance characteristics (7a–c), monitoring details
	h) QA, rejection criteria and rejection summary
	i) Maintenance findings: discussion of recommended maintenance schedules
8.	Conclusions, Performance Claims, and Limitations
9.	Third Party Review: The testing and reporting, if not performed by an independent professional third party, must be reviewed. The independent review should include a review summary and observation of at least one monitoring event
10.	Appendices: raw data and credentials

performance of stormwater management strategies. Continued population of the database and assessment of its data will lead to a better understanding of factors influencing BMP performance and help to promote improvements in stormwater control measure design, selection, and implementation. Statistical performance reporting requirements consistent with those used in the BMPDB are proposed as part of the MTD reporting framework. The BMPDB is only as valuable as the information that is included in it. The breadth and depth of the testing that will be done for MTD certification are ideal for inclusion in the BMPDB.

References

ASCE, U.S. Environmental Protection Agency, Water Environment Research Foundation, Federal Highway Administration, American Public Works Association, Wright Water Engineers, and Geosyntec Consultants. (1996). "International BMP database." <http://www.bmpdatabase.org/> (Jul. 12, 2010).

Ecology. (2008). "Technology assessment protocol—Ecology's (TAPE)." Dept. of Ecology, Washington, DC.

Lenhart, J. H. (2007). "BMP performance expectation functions—A simple method for evaluating stormwater treatment BMP performance data." *9th Biennial Conf. on Stormwater Research and Watershed Management*, Univ. of Central Florida, Orlando, FL.

NJCAT. (2007). "NJCAT technology verification, stormwater management Stormfilter." CONTECH Stormwater Solutions, Inc., Bordentown, NJ.

NJDEP (New Jersey Department of Environmental Protection). (2006). "New Jersey tier II stormwater test requirements—Amendments to tarp tier II protocol." New Jersey Dept. of Environmental Protection, Trenton, NJ.

TARP. (2003). "TARP (The Technology Acceptance Reciprocity Partnership) protocol for stormwater best management practice demonstrations." New Jersey Dept. of Environmental Protection, Trenton, NJ.

CHAPTER 6

Inspection and Maintenance

Maintenance Subcommittee

The recommendations for manufactured treatment device (MTD) maintenance contained in this chapter cover the following seven areas: (1) designing for maintenance, (2) defining standard maintenance triggers, (3) defining maintenance fundamentals for all MTDs, (4) defining maintenance tasks by MTD design and hydrodynamic or filter design, (5) identifying entities best able to maintain MTDs, (6) identifying entities to train maintenance providers, and (7) reviewing recommended disposal techniques for captured pollutants.

6.1 DESIGNING FOR MAINTENANCE

To more effectively maintain any stormwater practice, including MTDs, the designer must consider the maintenance needs in the design of the MTD.

The following specific design recommendations related to maintenance should be followed whenever possible:

1. Vacuum trucks for cleaning MTDs are heavy trucks. A roadway designed for truck loadings should be provided giving access (including any turnarounds) to an MTD.
2. The MTD must be structurally designed for vertical and horizontal truck loadings.
3. The MTD should be located within 5 m (15 ft) of the access road, because the boom of a typical vacuum truck can only reach 5 m (15 ft).
4. The access road should enable a vacuum truck to reach the MTD site under its own power.
5. There should be easy access to all internal components and spaces of an MTD to facilitate long-term cleaning, inspections, and repairs. Some chambers within the MTD are either not accessible or very difficult to access. Being able to visually inspect components of the MTD must be factored into the design (Fig. 6.1).
6. It is recommended that access manholes have a minimum nominal diameter of 75 cm to facilitate cleaning and confined space entry. All appurtenances,

Fig. 6.1. Manual access to every chamber of an MTD is critical
Source: Courtesy of Richard Gee, Montgomery County, MD

such as hoses and booms, as well as the maintenance personnel, should be able to pass through the opening. Filter boxes and certain hydrodynamic devices may require larger openings because some components may not fit through a 75-cm diameter hole. At a minimum, a manually or remotely operated high-pressure spray washer should be able to reach all device surfaces to remove collected detritus.

7. The MTDs on roadways should be located so that one lane of traffic can remain open during cleaning and maintenance operations (Fig. 6.2).

6.2 MAINTENANCE TRIGGERS

When the MTD is turned over to the property owner/manager, it must be essentially clean of construction debris. It is the responsibility of the installer or contractor to leave the MTD in a clean state. After a clean MTD has been accepted by the maintenance authority, inspections should be made quarterly for one year to determine the appropriate cleanout intervals.

Cleanout operations will be triggered by any of the following circumstances or combinations of them:

1. A regularly scheduled cleanout interval predetermined by the manufacturer.
2. Sediment accumulations reach the depth recommended by the manufacturer for cleaning. The appropriate depth of sediment determination should be facilitated by a mark or object placed in the MTD. This indication should be readily visible under low light conditions.

Fig. 6.2. Designers should find a location that provides relatively easy access
Source: Courtesy of Dr. Qizhong Guo, Rutgers University

3. In filter devices, the water drawdown time exceeds the drawdown time recommended by the manufacturer. An easily readable plaque should be placed inside the MTD indicating the recommended drawdown time.
4. In a filter MTD, the life expectancy of the filter medium reaches the time period recommended by the manufacturer. Filter media often function to remove pollutants based on more than just filtration (i.e., ion exchange, in which case maintenance requirements cannot be dictated by visual appearance); therefore, life expectancy and maintenance intervals should be preestablished to ensure performance.

In field work performed by Rutgers for the New Jersey Department of Transportation, it was observed that under average conditions, the time interval between cleanouts for an MTD is four years (Guo and Kim 2013a), and the time interval correlates well with traffic volume and area of impervious surface (Guo and Kim 2013b). This estimate is based on a maximum allowable sediment depth of two feet. However, it is still necessary to schedule regular inspection of the device at least once or twice a year, because activities in the watershed around the site can affect its performance; for example, new construction sites within the drainage area or very heavy storm events may cause the MTD to fill up faster.

The time intervals and expense of MTD cleaning can be reduced by providing upstream pretreatment of gross solids. Removal of pollutants by a pretreatment device, however, only shifts the burden of maintenance to a device further upstream. The total expense of maintaining a system of stormwater best management practices (BMPs) may or may not be reduced with pretreatment.

6.3 MAINTENANCE FUNDAMENTALS

There are three maintenance fundamentals common to most MTDs:

1. A maintenance entity should consider creating a database listing the stormwater BMPs, their locations, cleanout dates, masses removed (in both liquid and solid phase), and appropriate comments for a given jurisdiction. This database would be used to schedule cleanout events. Forms should be generated and given to maintenance personnel to document cleanout activities. Tracking cleanout activities ensures compliance with National Pollution Discharge Elimination System and Total Maximum Daily Load (TMDL) requirements and allocations. At least three separate data forms are recommended to collect data for the device: MTD information data form, inspection form, and maintenance form (Guo and Kim, 2013a). These three forms will be used to build a complete database and a maintenance forecasting and scheduling system.
2. The MTD manufacturer should provide an operations and maintenance manual giving thorough instructions for cleaning and maintaining the device.
3. The MTD cleaner should provide before and after maintenance photographs of the device, which can be included as part of the cleanout form and added into a maintenance database.

As recommended by the New Jersey Department of Environmental Protection's *NJ Stormwater Best Management Practices Manual* (2004), it is also recommended to take preventive measures that will avoid excessive trash, debris, and organic matter from reaching the device. Some of these recommended measures are providing trash receptacles around the MTD, installing pet waste stations where appropriate, providing as much vegetative cover as possible in the watershed, using pervious paving materials, and leaving pervious spaces between impervious areas.

6.4 HYDRODYNAMIC VERSUS FILTER-BASED MTDs

Hydrodynamic devices and filter devices require distinctly different cleaning operations. Hydrodynamic devices may be maintained by implementing a functionality inspection per the manufacturer's recommendation, vacuuming the sedimentation vaults, cleaning any screens, and removing accumulated oils and greases.

Maintenance tasks associated with filter devices for removal of pollutants from the overall structure are similar to those of hydrodynamic devices, but there is an added layer of maintenance activities associated with the filter media and the associated cartridges. Filter devices are maintained by implementing a functionality inspection per the manufacturer's recommendation, vacuuming accumulated sediment vaults, replacing or cleaning filter cartridges or filter media, and cleaning any drawdown pipes. Additional lifting equipment and confined space entry may be required to remove and/or replace filter components.

6.5 RESPONSIBLE PARTY TO CONDUCT MAINTENANCE

MTD maintenance and cleaning should be carried out by properly trained municipal personnel or maintenance contractors. *Properly trained* personnel are expected to be able to demonstrate proficiency in at least the following basic functions:

1. Understanding the operation of the MTD;
2. Vacuum truck capability;
3. Special requirements of a specific MTD;
4. Proper material disposal;
5. Certification by the state environmental protection department to deal with disposal similarly to what is required for disposal of septic waste, trash, and residual municipal sewage sludge; and
6. Documentation of cleanout operations with pictures submitted to regulator/owner and added into a database.

It is also strongly recommended that all trained personnel who conduct manufactured stormwater control measure inspection and maintenance have confined space entry training (Fig. 6.3).

Fig. 6.3. Maintenance personnel should have completed confined space entry training
Source: Courtesy of Scott Perry, Imbrium

6.6 TRAINING RESPONSIBILITY

MTD-specific training should be provided by the MTD manufacturer. In the absence of training firms, detailed maintenance manuals specific to the MTD must be provided by the manufacturer.

6.7 DISPOSAL OF WASTES

Due to the high variability of pollutant loadings within drainage basins, there exists a potential for high concentrations of various pollutants within stormwater BMPs. Therefore, it is recommended that all materials removed from an MTD be disposed of in a properly permitted landfill in accordance with applicable local or state guidelines (Fig. 6.4). A characterization study performed by Guo and Kim (2013a) observed that the bottom sediments in MTDs were not hazardous and could be disposed at municipal landfills. However, the presence of total petroleum hydrocarbons can be frequently found and, depending on the local regulations, may be classified as a regulated waste. Regulated metals concentrations in sediments appears to be infrequent (Lenhart and Rick 2000).

A water quality analysis of water vacuumed out of MTDs (Guo and Kim 2013a) showed that total suspended solids and the chemical oxygen demand of stormwater is higher than median municipal wastewater levels. This indicated that it was advisable to decant the MTD liquid to wastewater treatment facilities whenever technically and economically feasible.

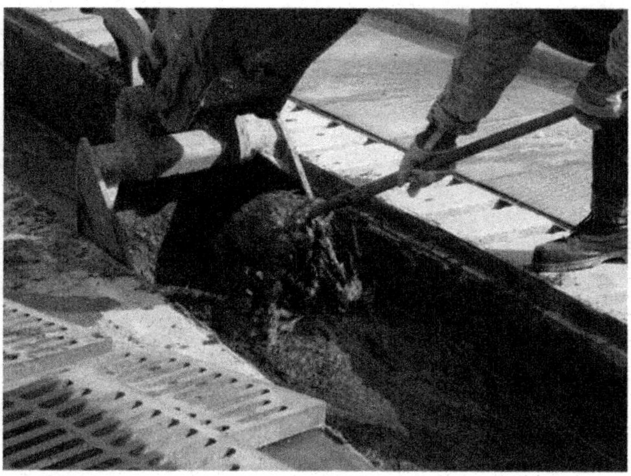

Fig. 6.4. Harvested materials from a Delaware-style sand filter
Source: Courtesy of William Hunt, NC State University

References

Guo, Q., and Kim, J. (2013a). "Stormwater system monitoring and evaluation." *FHWA-NJ-2014-10,* Federal Highway Administration, Washington, DC.

Guo, Q., and Kim, J. (2013b). "Stormwater system monitoring and evaluation—Implementation." *FHWA-NJ-2014-11,* Federal Highway Administration, Washington, DC.

Lenhart, J., and Rick, H. (2000). "Maintenance of stormwater quality treatment facilities." *American Public Works Association National Conf.,* American Public Works Association, Kansas City, MO.

NJDEP Division of Watershed Management. (2004). "NJ stormwater best management practices manual." <http://www.state.nj.us/dep/stormwater/bmp_manual2.htm> (Sep. 4, 2011)

CHAPTER 7

Existing Verification and Certification Programs and Processes

James Lenhart
Shirley Clark

The United States water pollution protection framework is a state-based regulatory system supported by national legislation (Clean Water Act 1972). Originally mandated by Presidential Order in the form of a permit program (the Refuse Act Permit Program), the subsequent amendments to this legislation regulated discharge of pollution into receiving waters without a permit. In 1972, the United States Environmental Protection Agency (USEPA) established the National Pollution Discharge Elimination System (NPDES) program to control pollution discharges and manage permits. Although the USEPA is the organization responsible for implementation of the NDPES, it may authorize states, territories, or tribes to implement all or parts of the program, including permitting and administration of the national pretreatment program, or administration of the municipal sewage sludge program. The USEPA maintains the right to review any permit issued on its behalf.

As a result of the NPDES system, several states have proactively developed legislation and regulations to manage emerging, innovative technologies aimed at the reduction of pollution into local receiving waters.

In 2001, the USEPA established the Environmental Technology Verification (ETV) Program. This program was closely followed by the development of state-based evaluation and verification programs. These included the Technology Assessment Protocol–Ecology (TAPE) Program and the Technology Acceptance Reciprocity Partnership (TARP) Program, which was overseen by the New Jersey Corporation for Advanced Technology (NJCAT). The ETV program was eventually ended, due largely to funding issues and that the outcome of the program did not provide specific approvals that open markets for product manufacturers.

There have been a number of other programs that have not gained as much national recognition, including the City of Sacramento, California; the State of North Carolina; and the Virginia Technology Assessment Protocol (VTAP) program in Virginia, which was never instigated. Other jurisdictions across the United States have specific criteria for gaining manufactured treatment device (MTD) approvals that include recognition of TAPE and TARP verifications and certifications.

More recently, the Water Environment Federation (WEF) has formed the Stormwater Testing and Evaluation for Products and Practices (STEPP) committee, which is looking into a nationally based program and protocol. A white paper (WEF 2014) was published, finding that a national program was both needed and feasible.

7.1 TAPE

The TAPE program was established by the Washington State Department of Ecology (ECOLOGY) in 2000 to address concerns that claims regarding best management practices (BMP) performance were not actually meeting the state's criteria for total suspended solids (TSS) removal (80%).

Having evolved over 16 years, several iterations of the program have been developed based on knowledge gained from earlier programs. Through state grants and participation fees, the Washington Stormwater Center (www.wastormwatercenter.org) was formed in 2009 to help facilitate the program for ECOLOGY. The Center also provides a complete array of services such as applied research, public outreach, and training. In addition, it provides a board of external reviewers that provide peer review of manufacturers' quality assurance project plans (QAPPs) and final technical evaluation reports. Once reviewed and deemed it meets the criteria and performance goals, a device is granted the appropriate use designation.

To participate in the program a manufacturer must apply for one of three use designations, each of which requires field testing that meets the TAPE protocol. A pilot level designation is assigned for laboratory scale–based evidence. At this level, the manufacturer is allowed to install five sites and provide field monitoring to demonstrate it meets the performance goals. The second level is the conditional use level designation (CULD). To gain CULD, the manufacturer must present both laboratory and substantive field data that may not meet the full TAPE protocol criteria. If granted, then 10 sites are allowed for field monitoring to demonstrate the technology meets the performance goals including preparation of a technical evaluation report (TER). Lastly, the general use level designation (GULD) allows for unlimited use of the technology assuming it meets both the protocol and the performance goals (Table 7.1). Once granted the GULD, ECOLOGY produces a certification letter that stipulates the specific design criteria, performance limits, and maintenance guidelines for the device.

The process includes preliminary laboratory analysis followed by submission of application documents to ECOLOGY (Fig. 7.1). This method provides a good level of independence and peer review to ensure confidence in the data and

Table 7.1. TAPE performance goals

Performance Goal	*Influent Range*	*Criteria*
Basic treatment (TSS)	20–100 mg/L TSS	Effluent goal ≤ 20 mg/L TSS
	100–200 mg/L TSS	Effluent goal ≤ 20 mg/L TSS
	>200 mg/L TSS	Effluent goal ≤ 20 mg/L TSS
Enhanced treatment (dissolved metals)	Dissolved Cu 0.005–0.02 mg/L	>30% dissolved Cu removal
Phosphorus treatment	Total Phosphorus (TP) 0.1–0.5 mg/L	Must meet basic treatment goal and exhibit ≥ 50% TP removal
Oil control	> 10 mg/L	1. No ongoing or recurring visible sheen in effluent
		2. Daily average effluent TPH concentration < 10 mg/L
		3. Maximum effluent TPH concentration of 15 mg/L or for a discrete sample
Pretreatment	< 50–100 mg/L TSS	≤ 50 mg/L TSS
	≥ 100 mg/L TSS	≥ 50% TSS removal

Source: Adapted from Table 2 in Washington Department of Ecology (2011a)

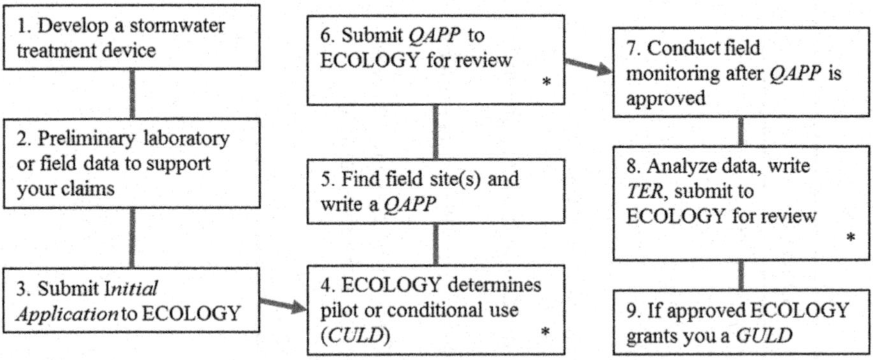

Fig. 7.1. U.S. stormwater quality improvement device (SQID) certification process

* Washington Stormwater Center and Board of External Reviewers, staged approval points
Source: Data from Washington Department of Ecology (2011b)

certification processes and reporting. It also seeks to establish minimum performance levels and an equal playing field for all technologies.

After the granting of a GULD, approved technologies are then posted on ECOLOGY's website (www.ecy.wa.gov/programs/wq/stormwater/newtech/technologies.html). Technologies approved by ECOLOGY can then be accepted by permitted jurisdictions. However, individual jurisdictions may choose not to accept the technology for use. The use of the product must also meet stated sizing requirements and designated maintenance requirements.

7.2 TARP

TARP was originally a consortium of six states that worked together to form the technology evaluation engineering report (Tier I and Tier II) protocols. Tier I focused on a CULD, using an established laboratory protocol and defined particle size distributions (PSDs) of surrogate sediment for hydrodynamic separators (HDSs) and filters. Tier II was a protocol for field testing to demonstrate a removal of 80% TSS for filters and 50% TSS removal for HDSs.

TARP was administered by the NJCAT (www.NJCAT.org). This government-legislated organization was tasked with the management and verification of these studies, among the verification of other technologies for energy and environmental remediation. Once a verification was completed, the manufacturer could approach each state for individual certification. This allows each state to assess the technology relative to its specific design requirements.

The consortium has since lost cohesion and continues to only serve the State of New Jersey, though other states and local jurisdictions may recognize the program. The protocol was subsequently revised and restricted to the acceptance of products via laboratory verification. Laboratory testing protocols were developed using specific PSDs (different than those used for TIER I) of silica sand for both HDSs and filters. HDSs can be used for pretreatment or standalone treatment, and filters can be used as standalone treatment. These protocols were developed in concert with the Stormwater Equipment Manufacturers Association.

The TARP protocol specifies that a manufacturer must present QAPP for review and approval. If accepted, the manufacturer moves to a test phase with an independent observer. If not accepted, changes are negotiated with NJCAT. If not resolved, an independent review is conducted by selected academic sources. Once the testing is complete a draft finding report is published by NJCAT. This report is placed for public comment, which includes other manufacturers. After the public exhibition period, changes are made if all comments are addressed and resolved. If not resolved then there is another opportunity for independent review. Once the report is finalized, it is issued as a verification and can then be used for certification by the State of New Jersey Department of Environmental Protection or other jurisdictions that recognize this process.

7.3 WEF

In 2012, WEF representatives, manufacturers, the USEPA, consultants, and other state and local jurisdiction staff met to discuss the issue of verification and certifications of manufactured products. Major points of discussion included the cost and time requirements of repeat testing required by many jurisdictions.

The STEPP working group was formed in 2013 to investigate the feasibility of a national program. The working group investigated a number of key challenges, including the development of appropriate program architecture to support adequate project management and peer review processes and ensure an objective and transparent approvals process. Regional test facilities were also identified as a future option to support rapid, controlled field testing and certification.

The working group also resolved certification processes that were to be undertaken by the national program and verification by individual agencies to certify the technology for use in their own jurisdictions. The certification outlines specific needs such as sizing criteria and maintenance requirements. There is also general consensus that this program should be extended to all stormwater control measures, not just manufactured ones. This would extend the scope of monitoring parameters to nutrients, bacteria, metals, runoff volume management, and even maintenance.

There is also a recognized need for uniform reporting and data management and information dissemination. Examples include using the International Stormwater BMP Database (www.bmpdatabase.org), which currently has data from about 530 studies. Funding is also seen as critical to a successful program. Sources of funding have been identified to include manufacturers, subscribers, grants, and training. The USEPA has already provided $100,000 in funding to move the process forward. The final report was published in July, 2016 (WERF 2016).

7.4 OTHER ORGANIZATIONS

Other organizations of note that are involved in MTD testing and verification include Urban Water Resources Research Council, which is in the process of finalizing a general guidance document on MTDs. The International Stormwater BMP Database, referenced previously, is also the collaborative effort of ASCE and the USEPA. In 2004, the project transitioned to a more broadly supported coalition of partners led by the Water Environment Research Foundation, including the Federal Highway Administration, American Public Works Association, and the Environmental and Water Resources Institute of ASCE.

ASTM International (ASTM) has formed a subcommittee on precast concrete products for stormwater management (ASTM C27.70), which provides methods and standards for testing stormwater quality improvement devices, mostly in a laboratory environment. Completed standards include test methods for the measurement of hydraulic characteristics of HDSs and underground settling devices (ASTM C1745M-11), a standard test method for suspended sediment removal efficiency of hydrodynamic stormwater separators and underground settling devices (ASTM C1746M-12), and a standard test method for measurement of hydraulic characteristics of stormwater filtration elements.

References

ASTM. (2011). "Standard test method for measurement of hydraulic characteristics of hydrodynamic stormwater separators and underground settling devices." *ASTM C1745/C1745M-11*, West Conshohocken, PA.

ASTM. (2012). "Standard test method for measurement of suspended sediment removal efficiency of hydrodynamic stormwater separators and underground settling devices." *ASTM C1746/C1746M-12*, West Conshohocken, PA.

ASTM. (2016). "Standard test method for measurement of hydraulic characteristics of stormwater filtration elements." *ASTM C1814/C1814M-16*, West Conshohocken, PA.

Clean Water Act. (1972). "Federal water pollution act 1972." *Congress of the United States*, Washington, DC.

Washington Department of Ecology. (2011a). "Technical guidance manual for evaluating emerging stormwater treatment technologies: Technology Assessment Protocol—Ecology (TAPE)." <https://fortress.wa.gov/ecy/publications/documents/1110061.pdf> (Apr. 26, 2017).

Washington Department of Ecology. (2011b). "Technology assessment protocol—Ecology (TAPE): Process overview." <https://fortress.wa.gov/ecy/publications/documents/1110010.pdf> (Apr. 26, 2017).

WEF (Water Environment Federation). (2014). "Investigation into the feasibility of a national testing and evaluation program for stormwater products and practices." <http://wefstormwaterinstitute.org/wp-content/uploads/2016/08/WEF-STEPP-white-paper_Final_02-06-142.pdf> (Aug. 10, 2017)

WERF (Water Environment and Reuse Foundation). (2016). "Framework for a national testing and evaluation program based upon the national Stormwater Testing and Evaluation for Products and Practices (STEPP) Initiative." Water Environment and Reuse Foundation, Alexandria, VA, *Final Rep. INFR2R14*.

CHAPTER 8

Conclusions and Recommendations

Qizhong Guo

Agencies are realizing that approval, control, and regulatory oversight of manufactured treatment devices (MTDs) is difficult without an MTD approval process and program. In addition, establishing a program is time consuming and expensive and requires multidisciplinary expertise.

An example is given where an agency allows a new MTD to be installed for field testing to evaluate total suspended solids (TSS) removal and sediment loading characteristics. Based on lab data, it takes six months to agree upon the conditions of monitoring, the manufacturer seeks a project for the test site (three months), the site goes to construction (eight months), construction activities (three months to one year), site stabilization and pollutant buildup (one year), testing (two years), and reporting and review (six months). The process could take years, and even then, the conclusion could be that the site was very clean and did not generate enough pollutants to determine if the MTD was meeting the goals.

This is why many agencies provide conditional or limited levels of approval based on the data presented. In many cases, limited laboratory testing data inhibits MTD approvals until long-term effectiveness at several local sites is demonstrated. Extensive field data and successful installations increase a regulator's willingness to grant conditional approvals for MTDs.

Consequently, many agencies seek to build on existing MTD verification programs by reviewing those programs within the context of the local conditions and regulations, establishing appropriate design guidelines, and providing some level of approval. Some existing programs have been developing for more than 10 years, with many improvements and lessons learned.

These Guidelines were crafted as a tool to assist communities with development of MTD evaluation and approval processes with respect to sediment and solids removal from stormwater runoff. Investigations and literature reviews were undertaken to evaluate many of the existing MTD verification programs. Drawing upon the experience and lessons learned from existing programs, this document identified seven major components that should be considered to establish a successful MTD evaluation and certification program:

1. Laboratory testing,
2. Scaling relationships,
3. Field testing,
4. Field data reporting,
5. Field data analysis and performance evaluation,
6. Maintenance requirements, and
7. Local conditions and requirements.

8.1 LABORATORY TESTING

Laboratory testing methods were developed for hydrodynamic and filtration-type MTDs. Major recommendations included the following.

8.1.1 Hydrodynamic Devices

- Use of total mass balance techniques to calculate a removal efficiency.
- Use of suspended sediment concentration (SSC) parameter rather than TSS parameter.
- Flow rates should be tested at 25, 50, 75, 100, and 125% of an MTD's maximum design treatment rate.
- Three ranges of particle sizes should be tested at each flow rate. The particle size distribution (PSD) of each range should be customized for the soil types in the region of approval.
- Influent sediment concentration should be 200 ± 40 mg/L of SSC.
- Testing for scour.

8.1.2 Filtration Devices

- Use of total mass balance techniques to calculate mass removal rates per surface area and mass removal efficiency.
- Calculation of mass loading capacity.
- Flow rates should be tested at 25, 50, 75, 100, and 125% of an MTD's maximum design treatment rate.
- Use of SSC parameter and PSD.
- Measurement of particle gradation at various surface loading rates.
- Measurement of head losses through a filter medium.
- Influent concentration should be between 100 and 200 mg/L of SSC.

8.2 SCALING RELATIONSHIPS

Verifiable scaling relationships have not been documented for hydrodynamic or filtration devices. Only full-scale testing of MTDs should be used.

8.3 FIELD TESTING

No recommendation is made for use of field testing to verify laboratory testing. Calculate removal efficiencies through event mean concentration measurements. A minimum 15 storm events should be sampled, with a preference for at least 10 aliquots collected per event. At least 80% of a given storm must be captured. SSC testing is recommended, but TSS and particulate matter evaluation are also accepted or encouraged. Sampled event mean flows should be reasonably distributed between 25 and 75% of the MTD design flow rate.

8.4 FIELD DATA REPORTING

Data should be reported in a format compliant with the International Stormwater Best Management Practices (BMP) Database requirements.

8.5 FIELD DATA ANALYSIS AND PERFORMANCE EVALUATION

A number of statistical analytical methods were discussed. While a community may use any of these methods, a Shapiro-Wilk W test appears to accurately determine the statistical validity of a dataset. Once the field data has been analyzed it should be evaluated and compared with a performance standard. Various methods are presented for evaluation of MTD performance. A recommended methodology is a *performance expectation function* that combines concentration removal efficiencies for high sediment concentrations and concentration predictions for low influent concentrations that approach an irreducible concentration level.

8.6 INSPECTION AND MAINTENANCE

Inspection and maintenance of MTDs is essential to ensure long-term pollutant removal effectiveness of the device. A review of design factors that facilitate proper maintenance of an MTD is given. Essentials of personnel training, maintenance procedures, and record keeping are also discussed.

8.7 VERIFICATION AND CERTIFICATION PROGRAMS AND PROCESSES

A review of existing and emerging verification and certification programs was undertaken. There are two components of an MTD program. The first is use of consistent, scientifically defendable methods to evaluate and verify stormwater control measure performance in general and specifically MTD performance. The International Stormwater BMP Database and several programs (such as the Technology Assessment Protocol–Ecology, Technology Acceptance Reciprocity Partnership, and New Jersey Corporation for Advanced Technology programs) have developed significant databases of MTD testing results.

The second component is a process that certifies that an MTD meets a stated or desired level of performance. There is no one best program. Each community should evaluate MTD performance goals based upon local environmental and total maximum daily load (TMDL) objectives. The goal of this committee is to provide a wide variety of program evaluations for a community to evaluate and possibly use for reciprocity.

Reciprocity provides a community with advantages of confidently using previous testing and verification from independent third parties, minimizing resources for starting and maintaining an MTD certification program, and minimizing conflict with MTD manufacturers. In turn, manufacturers would have a *level playing field* with consistent design guidelines and approvals, resulting in reduced costs and overhead that would be passed along to the communities.

To establish a successful MTD approval program, the following elements should be included as a minimum:

- Ensure the regulations are defined, achievable, and realistic. For example, a simple 80% TSS removal is not well defined and will lead to confusion and a breakdown of the process by which technologies are evaluated.
- Define intangible requirements such as maintenance needs and costs and climate issues (e.g., cold weather).
- Understand and document the specific MTD design requirements to meet the defined water quality and quantity requirements set within a jurisdiction. Provide these Guidelines to the design community to ensure proper design.
- Submit detailed designs to trained plan reviewers to ensure proper design.
- Re-review plans, calculations, specifications, and sizing criteria rigorously for both public and private development.
- Implement proper training of construction inspectors to ensure proper installation and inspection of completed system.
- Adequately budget for continuing education and training of staff.
- Document the goals and process in the manuals and keep MTD-specific information and approvals on websites or with letters of approval. Maintain the right to revoke or suspend approvals or change conditions of approval.

- Separate MTD approval processes from general stormwater design regulations. The reviewer should have the flexibility to adjust design requirements for site-specific conditions.
- Provide designers and vendors with easy access to MTD approval processes on websites.

The rapid growth and development of MTDs are driven both by engineering needs and economic conditions. There is no doubt that as regulations become more stringent and more pollutants are regulated, new technologies will emerge through innovation and continuous product improvement. The regulatory community should move forward to establish regulations and processes that provide a level playing field while keeping in mind the concurrent objective of protecting water resources.

Index

Page numbers followed by *e, f,* or *t* indicate equations, figures or tables.